Seventy Years of Farm Machinery

Part Two:
Harvest

Seventy Years of Farm Machinery

Part Two: Harvest

Brian Bell MBE

OLD POND PUBLISHING

IPSWICH

First published 2010

Copyright © Brian Bell, 2010

The moral right of the author in this work has been asserted

All rights reserved. No parts of this publication may be reproduced, stored in a retrieval system, or transmitted, in any form or by any means electronic, mechanical, photocopying, recording or otherwise, without prior permission of Old Pond Publishing.

ISBN 978-1-906853-50-1

A catalogue record for this book is available from the British Library

Published by
Old Pond Publishing Ltd
Dencora Business Centre, 36 White House Road
Ipswich IP1 5LT United Kingdom

www.oldpond.com

Cover design and book layout by Liz Whatling
Printed and bound in China
on behalf of Latitude Press

Contents

Conversion Table..6

Introduction...7

1 BINDERS AND COMBINE HARVESTERS
Reapers and binders – threshing machines
– stripper harvesters – windrowers – combine harvesters
– into the 1980s – pea and bean harvesters.......................................9

2 BALERS
Stationary balers and trussers – pick-up balers
– into the 1980s – big balers – bale handling...................................54

3 HAY AND SILAGE MACHINERY
Cutter bar mowers – mounted mowers
– mid-mounted mowers – rotary mowers – haymaking
– hay loaders – silage machinery – green crop loaders
– forage wagons – forage harvesters..86

4 SUGAR BEET MACHINERY
Down-the-row thinners – harvesters
– self-propelled sugar beet harvesters
– cleaner loaders..123

5 POTATO HARVESTING MACHINERY
Spinners and elevator diggers – complete harvesters
– haulm pulverisers...151

6 ESTATE MAINTENANCE MACHINERY
Hedging and ditching – post-hole diggers
– ditching machinery – saws – chainsaws
– lifting and carrying – farm trailers..177

Index...211

Conversion Table

The following information is offered to help those readers who may be too young to remember the imperial days of pounds, shillings and pence, acres, hundredweights and coombs.

1 gallon = 4.6 litres

1 inch = 25.4 mm
1 foot = 305 mm
1 yard = 910 mm
1 chain = 22 yds
1 acre = 0.4 hectare
2½ acres = 1 hectare

1cwt = 50.8 kg
1 ton = 1,016 kg

1 shilling = 5 pence
£1 = 20 shillings

Bushel and coomb are volumetric measures, the weight of which will vary according to the density of the crop.   There are four bushels in a coomb.

1 coomb of wheat weighss 248 lb = 2¼ cwt

1 coomb of barley weighs 224 lb = 2 cwt

1 coomb of oats weighs 168 lb = 1½ cwt

Introduction

During the twentieth century there was not only a revolution in the design of farm machinery for sowing and growing arable and forage crops but also changes beyond recognition in the machinery used to harvest cereals, grass and root crops.

Average wheat yields in the 1930s rarely exceeded one imperial ton an acre, or little more than two metric tonnes per hectare. At that time the grass and cereal harvest depended to a great extent on hand labour. Reciprocating knife cutter bar mowers and tedders had mechanised the hay harvest and cereal crops were cut and tied in sheaves with a self-binder. However, a great deal of manual labour was still needed at haysel and harvest time. Loose hay was loaded and carted to the haystack and, after stooking, the sheaves were left in the field for three Sundays before being carted to the stack yard.

Sugar beet was already grown in Holland, Germany and neighbouring countries by the mid-1800s. After a failed attempt to grow and process the sugar beet crop in Britain in the 1830s another eighty years passed before the first successful British sugar factory was opened, in Norfolk in 1912. There were eighteen sugar factories by 1928 when the entire sugar beet crop was lifted, knocked and topped by hand.

There was at least one tractor on most farms with 100 acres or more in 1939 when working horses outnumbered the total population of no more than 55,000 tractors by a ratio of thirteen to one. Combine harvesters were few and far between but most farmers owned a self-binder. A contractor with a traction engine and drum would be hired to thresh the stacks of corn. By 1950 there were at least 250,000 tractors on UK farms but even then working horses still outnumbered tractors by a ratio of two to one.

The farm workforce gradually reduced over the years. One man was employed for every 25 acres of arable land in the early 1900s but the labour force had dropped to one worker for 75 acres by 1950. By 2000 some farms of 800 acres or more with powerful four-wheel drive tractors and self-propelled harvesting machinery employed just one highly skilled full-time farm worker.

Part One of *Seventy Years of Farm Machinery - Seed Time* presented an overview of the development of ploughs, cultivators, drills and crop care machinery. Part Two charts the development of harvest machinery for cereals, hay, silage, sugar beet and potatoes. In the short space of seventy years sail reapers, binders and threshing machines have given way to the combine harvester. Self-propelled foragers and 24-hour haymaking techniques have replaced the cutter bar mower, hay loader and buckrake. Harvesters now do in a single pass the once labour-intensive task of securing potatoes and sugar beet.

Brian Bell
Suffolk, 2010

Chapter 1
Binders and Combine Harvesters

Although combine harvesters were well established in America by the mid-1940s very few were in use on British farms where the self-binder and threshing drum still dominated the harvest fields. Tractors had replaced the traction engine as the power source for threshing tackle but horses were still hauling wagons loaded with sheaves from the harvest field to the stack yard. The UK combine population had grown by the early 1950s. Some models were self-propelled but most were towed by a tractor and either driven by a separate engine or power take-off (pto) shaft.

Although many farmers were using a combine harvester in the early 1950s Albion, Bisset, International Harvester, Massey-Harris and others were still making a limited number of binders. The agricultural machinery census for England and Wales in 1958 recorded a total of 99,000 binders compared with 40,000 combine harvesters. However, these figures did not indicate how many binders were actually in use.

Reapers and Binders

An Arbroath parson, Rev Patrick Bell, built one of the first reaping machines in 1826. Pushed by a horse, Bell's reaper had a reciprocating knife with double-edged blades to cut the crop and a cross conveyor to carry it sideways and deposit it in a swath on the stubble. Cyrus Hall McCormick began making a similar horse-drawn reaping machine in America in 1831. Two men were needed for the McCormick reaper, one steered the horse and the second, who walked behind the reaper, raked bunches of the crop sideways off the platform, leaving them on the stubble to be tied by hand. A later version had a seat for the rake man.

The introduction of the automatic self-rake or sail reaper, patented by McCormick in 1862, made the

1.1 The McCormick Daisy sail-reaper with a folding cutter bar platform was made in America.

rake man redundant. Vast numbers of the McCormick Old Reliable sail reaper, made until 1875, were sold in America and beyond. The folding McCormick Daisy reaper, made between 1890 and 1905, was an improvement on the earlier models as the sails only swept the crop off the platform when there was enough to make and tie a sheaf. The last Daisy reapers were made in 1934.

The Marsh reaper, first made in America in 1862, marked the next stage in the mechanisation of the grain harvest. A canvas conveyor with a mechanical bunching mechanism gathered the crop in loose bundles and two men riding on the machine tied the bundles with straw bonds. The sheaves were dropped off in heaps at intervals across the field. John Appleby, an Englishman living in America, took out a patent in 1869 for a mechanical wire-twisting mechanism to bind sheaves. Walter A Wood was one of several early 1870s American companies using wire twisters for their reaping machines. Wire twisters were also used on the 50,000 or so McCormick reapers made between 1877 and 1885. The wire twister worked well enough but livestock farmers feared that their animals might be harmed by odd pieces of wire in the straw.

The birth of the twine-tying reaper binder dates back to 1879 when John Appleby perfected his knotter. The Deering Harvester Co, the Minneapolis Harvester Company and others were using the Appleby knotter for their reaper binders in 1880 and McCormick obtained a licence to use the knotter in 1881. Within three years McCormick had sold 15,000 twine-tying reaper binders with Appleby knotters. To ensure an adequate supply of binder twine the company opened a five-storey twine mill in the early 1880s. Fifty years later the 1,500 men and women employed at the McCormick twine mill in Chicago were making up to 30,000 tons of twine a year.

Ground-wheel driven binders, hauled by teams of two, three or four horses, were a common sight in the early 1900s but in very wet areas where ground-wheel drive did not work some farmers used a binder with a petrol engine instead. Horse binders worked at speeds of around 2½ mph but that dropped down to 2 mph as the team became tired. Two teams were needed for a full day's work with a binder as the working limit for one team of horses pulling a 4 ft cut binder was no more than five to six hours a day.

1.2 Albion No 3 horse-drawn binders were made between 1904 and 1922.

Some binders had a sheaf-carrier below the knotter deck. This collected the sheaves and dropped them in heaps at intervals across the field to help speed up the stooking, or shocking up, process. Stooking became mechanised in 1914 when International Harvester introduced an attachment for the job. The complicated mechanism, which formed the sheaves into stooks and left them in rows across the field, enjoyed minimal success.

Harvest with a binder started about three weeks earlier than it does with a combine harvester. It was usual to mow a strip around the field headland with a scythe to clear a space for the tractor wheels when cutting the first round with the binder. The hand-mown corn was gathered into sheaves and tied with bands made by twisting together some lengths of straw.

Unlike modern farm machinery, which is updated every two or three years, designs lasted much longer in the early twentieth century. The ground-drive Albion No 3 left- and right-hand cut binders with a draught pole for horses or a tractor drawbar introduced by Harrison, McGregor & Guest in 1904 remained virtually unchanged for eighteen years. Chain and gear drives were used throughout and the instruction manual advised that all bearings should be lubricated with good oil. It was stressed that common oils should not be used.

Towing the binder from one field to another involved re-positioning the drawbar or horse pole from the front to the cutter bar end of the binder and attaching transport wheels in front of and behind the knotter deck. The main driving (or bull) wheel, used to adjust cutting height, provided a built-in jacking system when changing from field to road position or vice versa.

Tractors pulled ground-wheel drive binders twice as fast as a team of horses and they didn't need to stop for a rest. Some farmers used the services of a local blacksmith to make a tractor drawbar for their horse binder but horse binders were not built to work at tractor speed. More robust tractor-drawn wheel-driven binders with wider cutter bars were on the market by 1910. The first 5 and 6 ft cut pto-driven binders appeared in the early 1920s and within ten years pneumatic tyres were an optional extra for some power-driven binders.

Binders cut the crop with a smooth sectioned knife and the reel guided the crop on to an endless canvas cross conveyor on the cutter bar platform. The platform canvas carried the crop sideways to the lower and upper canvas elevators which transported it to the knotter deck. Here the packer fingers formed a sheaf against a single twine running from the needle eye to the knotter. When the sheaf reached the required size the drive to the knotter mechanism was engaged and the needle entered the knotter. The tied sheaf was thrown to the ground. A reciprocating buttor board kept the butt or bottom of the sheaf level while it was being made.

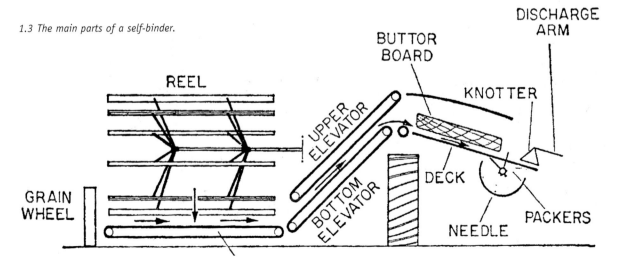

1.3 The main parts of a self-binder.

1.4 A 6 ft 6 in cut Lanz binder made in Germany in 1946 cutting wheat for thatching straw.

The man on the binder seat was kept busy adjusting the height of the cutter bar and the position of the reel. He also had to engage and disengage the drive at the headlands and check that the twine was correctly positioned on the sheaf. The binder needed a great deal of maintenance with grease nipples and oil holes requiring frequent attention. Two knives were supplied with the binder and regular sharpening with a hand file was an essential chore. Some farmers used a specially shaped hand-operated grinding wheel to sharpen their binder and cutter bar mower knives.

The next twenty years saw few changes in binder design. With a growing number of

1.5 McCormick F-7-T power-drive binders were made in France.

combine harvesters on farms the days of the binder were clearly drawing to a close although Albion, Bisset, McCormick International, Massey-Harris and Wallace were still making them in the mid-1950s. McCormick binders included the No 6R ground-drive model and the pto-driven 5T, 6T and 7T, the number denoting the cutting width. The French-built 5 and 6 ft cut F-7-R ground-drive binders and the power-driven 7 ft cut F-7-T announced in 1954 were the last of a McCormick family of binders that spanned a period of more than 70 years.

The 5 ft and 6 ft cut pto- and ground wheel-driven Albion 5A binders were introduced in the early 1940s and the last ones made in 1958, some three years after Harrison, McGregor & Guest became part of the David Brown organisation. Available for horse or tractor power, standard equipment for the 5, 6 and 7 ft cut Albion 5A included steel or pneumatic-tyred wheels, a waterproof cover and a sheaf carrier. Instead of the binder dropping single sheaves in rows across the field, the sheaf carrier saved labour by collecting a number of sheaves and dropping them in heaps for stooking.

1.6 J Bisset & Sons were making 5 and 6 ft cut ground wheel-drive binders at Blairgowrie in the 1950s.

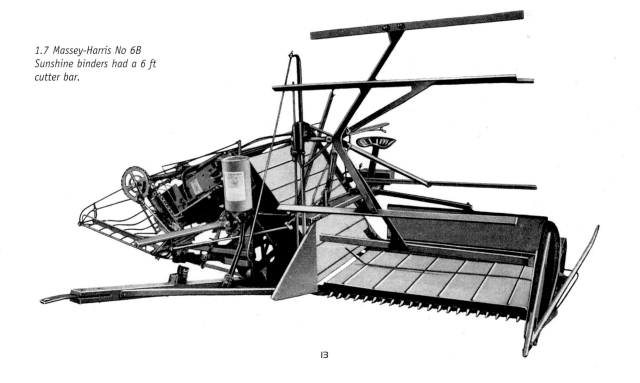

1.7 Massey-Harris No 6B Sunshine binders had a 6 ft cutter bar.

1.8 The Wallace JF Lightbinder introduced in 1955 had a single canvas and twin-pronged forks to pack the sheaves.

Massey-Harris, which became Massey-Harris-Ferguson in 1955, continued to make 6 ft left-hand cut Sunshine No 6B ground or pto-driven binders. Like other binders of the day they had a pressure lubrication system and the drive trains ran in oil baths.

Sales literature noted that the reel on the Sunshine binder could be set in any of 70 different positions to suit every possible crop condition. John Wallace & Sons at Glasgow introduced a rather different binder in 1955 in the shape of the Danish JF Lightbinder. It was unusual in that a single canvas conveyor carried the crop to a pair of twin-pronged forks used to form the sheaves before they were tied and thrown to the ground. J Bisset & Son at Blairgowrie in Perthshire, another long established binder manufacturer, was still making 6 ft cut ground wheel-drive tractor binders when they introduced 5 and 6 ft cut semi-mounted Bisset binders in 1950. Apart from the chain-driven knotter deck, silent vee-belt and shaft drives were used throughout the machine. The front of the binder was attached to the hydraulic lift arms and two castor wheels carried the back of the machine. It only took two minutes to fold up the platform and prepare the binder for transport, and with its pneumatic tyres the binder could be towed at the top speed of the tractor.

However, the last days of the binder were nigh and the one-man, semi-mounted Bisset with a £338 price tag was the only binder exhibited at the 1960 Royal Smithfield Show.

Threshing Machines

Hand flails were used to thresh grain until the 1840s when the first hand- and later steam-powered threshing machines came into use. Traction engines were used to haul and drive threshing machines in the 1860s and by the early 1900s Fisher Humphries, Foster, Marshall, Ransomes and Tullos were among the companies making rasp bar-type threshing machines. Ransomes made single- and double-blast machines with threshing cylinders from 36 to 60 in wide in the 1920s and 1930s. Blast referred to the type and output of the fan used to clear the chaff from the grain sieves. A typical mid-1930s contractor's thresher was a double-blast machine with a 54 in wide threshing cylinder. A double-blast threshing machine with a 48 in wide cylinder was considered adequate for the farmers who also did some contracting work to increase their income.

The threshing machine mechanism consisted of a cylinder and concave fed with crop by a man standing on top of the machine. From that point it worked in much the same way as most combine harvesters. Threshed grain fell through the concave grate on to the sieves and the straw walkers shook out the remaining grain from the straw as it made its way to the back of the machine. A fan blew the chaff off the sieves and the grain was elevated to the bagging-off spouts. The straw passed from the back of the machine on to an elevator which carried it up to two or three men

1.9 Ransomes' heavy-duty A60 thrashers with a 22 in diameter and 60 in wide cylinder had an output of up to 3½ tons an hour when threshing wheat.

who stacked it. On some farms the straw was fed into a belt-driven stationary baler or a chaff cutter.

More sophisticated threshing machines passed the grain to the awner and chobber and/or a rotary screen before it was bagged up in coomb sacks. When barley was threshed the grain was put through the awner and chobber where small spirally mounted knives removed the awns before it received a final rubbing and polishing in the chobber unit. The rotary screen, also used on some bagger combines, separated out the weed seeds and small grains from the good grain. The rotary screen consisted of a cylinder formed with steel wires. The space between the wires was narrow at the end where the grain entered the screen, the gap gradually increasing as the crop moved towards the bagging-off spouts for the best grain. The weed seeds and thin grains fell through the narrowly spaced wires while the seconds and best grain fell through wider spacings nearer the end of the cylinder.

1.10 Steam threshing with a Ransomes drum.

There were very few changes in threshing machine design over the years. The option of pneumatic-tyred wheels was an advantage for threshing contractors. Although more and more combine harvesters were coming into use a limited number of threshing machines, including the Marshall SM and Ransomes AM, were still made in the mid-1950s.

Stripper Harvesters

Dating back to the 1840s, stripper harvesters, which threshed grain from standing crops, were first seen in Australia and America. Although not first in the field, HV McKay of Sunshine, Victoria made Sunshine stripper harvesters in the early 1850s. Massey-Harris also experimented with stripper harvesters and by the early 1900s they were selling Canadian-built machines to Australian farmers. HV McKay became part of the Massey-Harris organisation in 1930.

With combine harvesters growing in popularity there was limited interest in stripper harvesters. However they re-appeared in 1946 when the Wild harvest thresher, a stripper under another name, was introduced to British farmers. The McConnel-Bomford-designed Wild harvest thresher Model 50 guided the standing crop between long triangular tines to an engine-driven threshing rotor running at 700 rpm. This removed the grain from the ears of the standing crop and passed it through an awner at the back of the machine where a man, riding on a platform, bagged up the threshed grain. A blower returned the chaff through two large diameter pipes to the ground. An air-cooled Petter engine was used to drive the threshing rotor, awner and blower. An advertisement explained that at a speed of 2½ mph and with two land girls working on the Model 50 the Wild harvest thresher would thresh about 8 acres of grain in a day.

The tractor-mounted Wild Model 40 harvest thresher, exhibited by MB Wild at the 1951 Royal Smithfield Show, was belt driven from the pto. Working in the same way as the Model 50 the long tines, situated alongside the rear tractor wheel, guided the ears to the threshing rotor. The threshed grain was bagged up on a platform at the back of the tractor.

1.11 The Wild harvest thresher.

Further developments in the mid-1980s resulted in the introduction of the Shelbourne Reynolds stripper header in time for the 1988 harvest season. The SRE stripper header, which replaced the conventional combine harvester cutter bar, had eight rows of keyhole-shaped teeth on a rotor running in an anti-clockwise direction. The teeth removed the ears and loose flag leaves from the standing crop which were carried by an auger and elevator to the threshing cylinder. Like earlier stripper harvesters the straw was left standing to be chopped, or harvested with a mower and baler. Although the SRE stripper header enjoyed little success in the UK models up to 32 ft wide were used on the sun-drenched Australian wheat fields in the early twenty-first century.

1.12 Shelbourne Reynolds stripper header was designed to replace a conventional combine cutter bar.

A prototype trailed stripper harvester was exhibited by Kidd Farm Machinery at the 1990 Royal Smithfield Show. Equipped with a Shelbourne Reynolds header the prototype had a threshing cylinder and concave, sieves and a fan. Four rotary separators replaced the usual straw walkers. It was claimed that the Kidd stripper harvester had an output equal to that of a medium-sized combine but was only half the price.

Windrowers

Although intended mainly for harvesting vining peas grown for freezing, a few companies made windrowers for cutting and windrowing other crops. The early 1950s trailed Albion No 10 windrower made by Harrison, McGregor & Guest was advertised as a machine for cutting and conditioning cereal crops on the stubble prior to combine harvesting. The No 10 windrower had a land wheel-driven reciprocating knife cutter bar and reel. The cut crop was carried sideways on a canvas conveyor and left in a swath on the stubble.

Early 1960s swather windrowers, including the Hume, JF, McBain and Reco (page 49), were used mainly to cut and windrow various green crops including oilseed rape and peas. Some farmers used a swather windrower to cut and swath lodged and wet

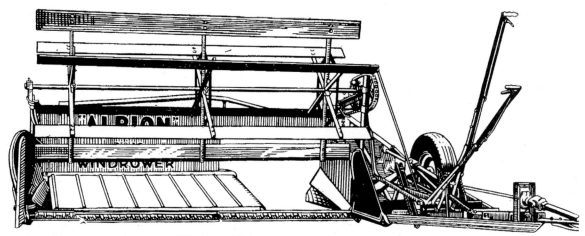

1.13 The Albion No 10 windrower cost £259 10s in 1954.

cereal crops. After leaving them to dry they were threshed with a combine harvester equipped with a pick-up cylinder.

Combine Harvesters

Large teams of horses or mules hauled the first combine harvesters, known at the time as reaper threshers or harvest threshers, across the Californian wheat belt in the late 1800s. Daniel Best made a self-propelled 50 ft cut steam-powered reaper thresher in 1893 and by the early 1900s animal- and steam-powered peg drum reaper threshers were harvesting much of the Californian wheat crop. The peg drum, which tended to damage the straw, dates back to the late 1780s. The rasp bar threshing mechanism, which did less damage, appeared in the early 1800s.

Teams of up to forty horses or mules were used in the late 1880s to haul the Holt reaper thresher across the American prairies. Benjamin Holt, who made the first self-levelling reaper thresher for working on steep hillsides in 1892, sold his first gasoline-engined reaper thresher in 1902 and built the first all-steel reaper thresher in 1913. Holt and Best joined forces in 1925 to form the Caterpillar Tractor Co. By 1929 they were making three models of trailed and engine-driven peg drum combine harvesters with a grain tank and a 10-20 ft wide cutter bar.

The first Deering bagger peg drum harvest thresher appeared in 1915. Eight or more horses or a tractor, if the farmer owned one, were needed to haul the 9 ft cut Deering No 1 harvest thresher. An auxiliary engine was available at extra cost. McCormick introduced the No 7 harvest thresher with a self-levelling 12 or 14 ft cutter bar for hillside work in 1926 and by the late 1930s they were making 6 ft cut, pto-driven trailed combine harvesters. The 12 ft cut No 31 harvest thresher, introduced in 1932, also had a peg drum. However, by 1941 all McCormick Deering combine harvesters, including the pto-driven No 42 and No 62, had a rasp bar threshing drum. The No 62, with a 12 cwt grain tank and optional Continental petrol engine, was advertised as being the ideally matched harvester for Farmall H and M tractors. During a ten-year

1.14 The cutter bar on the Claas Mah-Dresch-Binder could be detached and towed on a trailer behind the combine.

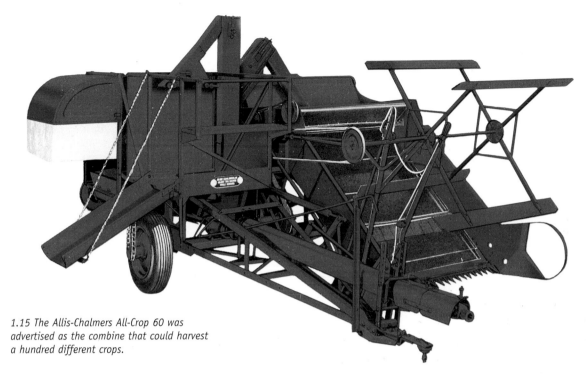

1.15 The Allis-Chalmers All-Crop 60 was advertised as the combine that could harvest a hundred different crops.

production run more than 40,000 International Harvester No 62 combines were made, with a good number of them exported to the UK.

A Massey-Harris reaper thresher from Canada and a 10 or 12 ft cut pto-driven McCormick Deering No 8 harvest thresher imported from America in 1928, were the first North American machines seen in the UK. Four men were needed for the engine-driven Massey-Harris. One drove the tractor, another sat at the back of the cutter bar platform to control the height of the reel and the cutter bar and the other two bagged off the grain.

A gang of four men, performing similar roles, was also needed for the first British-built combine harvester made by Clayton & Shuttleworth at Lincoln in 1929. Marshalls at Gainsborough bought the combine harvester division of Clayton & Shuttleworth in 1930.

In Germany, the first Claas combine harvester, wrapped around a Lanz Bulldog tractor, was put to work in 1930. It had a front cutter bar, a conveyor alongside the tractor to carry the crop to the threshing mechanisms and a bagging-off platform at the rear. Having failed to impress German farmers with the wrap-around combine, August Claas introduced the trailed Mah-Dresch-Binder combine harvester in 1937. The trailed M-D-B, which roughly translates as cut-thresh-tie, was a straight-through machine with the threshing drum and straw walkers in line with the cutter bar. The grain was bagged off at the rear and a trusser on the opposite side to the cutter bar tied the straw in bundles with single twine. The chaff was blown through a large diameter pipe to the ground but farmers wishing to collect the chaff could connect the chaff pipe to an enclosed trailer towed behind the machine. After replacing the cutter bar with a canvas conveyor, the combine could also be used to thresh sheaves in the field or at a stack. The last M-D-B combine harvesters were made in 1943 when the Claas factory was turned over to armament production.

Allis-Chalmers, Case, John Deere, Deering, Massey-Harris, McCormick, Minneapolis Moline and Oliver all made reaper threshers in America in the 1920s and 1930s.

The 5 ft cut Allis-Chalmers 60 trailed combine introduced in 1935 and later renamed the All-Crop 60 was made at Milwaukee until 1952. A few All-Crop 60 combines were in use on UK farms in the late 1930s and more were shipped to Britain in time for the 1941 harvest. The 4 ft cut All-Crop 40 made at Milwaukee for the smaller farm appeared in 1938. Allis-Chalmers All-Crop 60 and All-Crop 40 combines had a 15 in diameter threshing drum with rubber facings on the cylinder and concave rasp bars. The All-Crop 60 was one of the first pto-driven combines and from 1940 an Allis-Chalmers petrol engine was included in the list of optional extras. Another special feature was the combine's variable speed vee-belt drive to the drum. There were three speed ranges and three speed variations in each range adjusted with a crank handle linked to a variable diameter vee-belt pulley. The drum was at right angles to the cutter bar and after shaking out the grain the straw was left in a swath at the side of the combine. Following Allis-Chalmers' acquisition of the Minneapolis Moline factory at Essendine in 1950, the pto-driven All-Crop 60 with the option of a four-cylinder petrol or tvo engine was made there for ten years.

Case, which made their first combine harvesters in 1923, introduced the engine-driven Case Model Q in 1930. It followed the design of earlier American reaper threshers with a 12, 16 or 20 ft cutter bar, 32 in wide peg drum and 1½ ton gravity-emptied grain tank. The Model Q, along with a range of Case tractors, was marketed in the UK by Associated Manufacturers Co at the Palace of Industry in Wembley. Optional equipment for the Model Q included a bagging-off platform, a hitch for eight or twelve horses, pneumatic-tyred wheels and a grain meter that measured the flow of grain in pecks and bushels from the main elevator into the grain tank.

Early petrol engine-driven Massey-Harris peg drum reaper threshers include the horse-drawn No 1 introduced in 1910 and the No 5 launched in 1922. By 1937 Massey-Harris was making pto-driven No 15 and No 17 reaper threshers with rasp bar threshing drums. The pto-driven 5 ft cut trailed M-H Clipper and the 16 ft cut Massey-Harris No 20 appeared in 1938. The No 20 with a 37 in wide drum and six-cylinder 65 hp Chrysler engine, which made it rather top heavy, was one of the first self-propelled combine harvesters.

The better-balanced Massey-Harris No 21 with a Chrysler engine, a 12 ft cutter bar, a 32 in wide drum and a grain tank or bagging-off platform, superseded the No 20 in 1940. About 1,500 M-H21 combines were shipped from America to the UK in wooden crates between 1941 and 1948 to help with Britain's

1.16 The McCormick Deering 31T reaper thresher with a 12 or 15 ft wide cutter bar and a peg drum threshing cylinder was made in American between 1932 and 1940.

food production programme. The combines were delivered in their crates to farmers who, aided by dealer staff, put them together on the farm. Massey-Harris launched the 8 ft cut No 22 combine in 1944. Designed for the smaller farm, the No 22 with a 35 hp petrol engine was made with a grain tank or a bagging-off platform.

Combine harvesters, with the exception of the fleet of M-H21s, were quite a rare sight on British farms even in the late 1940s when most were owned by agricultural contractors. In answer to a parliamentary question the Minister of Agriculture stated that in spite of a shortage of materials it was expected there would be 5,000 combines in use for the 1947 harvest compared with 3,800 in 1946.

To celebrate 100 years of progress in farm machinery Massey-Harris announced in 1947 that it would again lead the field with the new 8 ft cut No 222 self-propelled combine harvester. It had a change-on-the-move 24 speed transmission, electric table lift and the choice of a 30 bushel grain tank or a four-spout bagger platform. Awarded a silver medal at the 1947 Royal Show in Lincolnshire, the No 222 was advertised as narrow enough to pass through a 9 ft gateway. However, little more was heard of the M-H222 and, along with the ageing No 21, it was replaced in 1949 by the 8 ft 6 in cut Massey-Harris 726. The new combine had a 2 ft wide belt-driven threshing drum and four straw walkers. Unlike earlier M-H combines with a manually raised cutter bar the table was lifted by an electric motor. Once again there was a choice of a bagging-off platform or a grain tank for the M-H726. Other optional equipment included a Raussendorf straw press, a straw spreader, a pick-up reel and a pick-up attachment for previously windrowed crops.

Massey-Harris, a company founded in Toronto in 1908, did not manufacture farm equipment in the UK until it opened a factory at Trafford Park in Manchester in 1945 but by the early 1950s Massey-Harris combines were a dominant force on British farms. Trailed and engine-driven versions of the 5 ft 6 in cut M-H750 for small acreage farms were launched at the 1952 Royal Show. Made at Kilmarnock, the M-H750 was a straight-through combine with a 5 ft wide drum, straw rack, grain sieves and a bagging platform complete with a rotary cleaning screen.

1.17 Massey-Harris No 21 combine harvesters had a 12 ft wide cutter bar. The platform canvas conveyor was replaced with a cable auger on the later M-H21A.

J Mann & Son at Saxham in Suffolk imported the first Claas Super trailed combines in 1947. By the early 1950s Allis-Chalmers, Albion, International Harvester, John Deere, Marshall, Massey-Harris, Minneapolis Moline and other trailed combines were at work in the UK.

Minneapolis Moline made the first MM Harvestor trailed combine harvesters in America in 1934. Several Harvestor combines were sent to Great Britain under the World War Two Lend-Lease programme. George Sale and Robert Tilney established Sale Tilney at Wokingham in 1938 to import Minneapolis Moline tractors and Harvestor combines from America. They also sold the 5 ft cut MM69 combine harvester in the UK. Introduced in 1940 and designed for small acreage farms, the MM69 was a straight-through combine with a 17 hp Wisconsin air-cooled petrol engine, a 69 in cutter bar, a 48 in wide drum and similar width straw rack and sieves. An unusual feature of the MM69 was alternate short and long fingers on the cutter bar which, it was claimed, would do a better job in damp or laid crops.

A new company trading as Minneapolis Moline (UK) took over the Sale Tilney premises at Wokingham in 1946. The first MM G4 and G8 tanker and bagger combines, based on the earlier MM G2, were made under licence in Wales in the same year. Available with a 30 hp Ford V8 or a 28 hp four-cylinder Meadows engine the 8 ft cut G8 had a 31 in wide drum. Production of Minneapolis Moline combine harvesters was transferred to Essendine in Lincolnshire. The G8 was discontinued in 1950 and Allis-Chalmers bought the Essendine factory in the same year. The American-designed Allis-Chalmers All-Crop 60 combine harvester was built there for the next ten years.

Sale Tilney also imported the engine-driven German-built Dechentrieter 6 ft cut combine harvester at Wokingham. Launched at the 1951 Royal Smithfield Show the first Dechentrieters were delivered in time for the 1952 harvest. With an output of up to 30 cwt an hour the combine, advertised as being equally suitable for threshing in the field or at the stack, was a straight-through machine with a two-band straw trusser.

Marshalls at Gainsborough, a firm making threshing machines in the 1850s, introduced a prototype 8 ft cut self-propelled combine in 1943. It had two 22 hp Ford petrol engines mounted on top of the combine. One was used to drive the cutter bar, 30 in wide drum and threshing mechanism and the second engine propelled the combine. Following modifications, which included replacing the Ford engines with two Coventry Climax power units, the Grain Marshall E9 Silver Queen was launched in 1945. As well as being

1.18 Minneapolis Moline made the 5 ft cut MM69 Harvestor in America.

1.19 The mid-1950s Dechentrieter combine harvester with an air-cooled petrol engine cost £1,295 ex works.

expensive to run it was considered short on manoeuvrability so Marshalls introduced a single tricycle wheel at the rear. This improved cornering but the combine was less stable and less than fifty had been sold when the Silver Queen was discontinued in 1947.

An advertisement in May 1949 for the new trailed Grain Marshall 568 combine, which cost £750 ex works, explained that delivery would be in strict rotation. The 5ft 6 in cut 568 had a wide straw rack and the reel was chain driven from twin wheels running on the stubble side of the combine. The threshing unit was driven by a 17 hp petrol engine. Variants of the 568 included the Grain Marshall 560 and the Grain Marshall 602. The 5 ft 6 in cut 602 had a 5 ft wide threshing drum with a similar width straw rack and grain sieves and a side-mounted straw trusser.

With its orange paintwork, the Marshall 626, launched at the 1953 Royal Show, was one of the most expensive engine-driven, trailed combine harvesters on the market. The 6 ft cut 626 combine, with an output of about an acre an hour, had a 25 hp Ford engine, a 36 in wide drum and four straw walkers. Optional equipment included a two-string Welger press, a straw spreader and a pick-up attachment for harvesting windrowed crops. Marshalls demonstrated a self-propelled prototype version of the Grain Marshall 626 with two engines in 1955 but thanks to strong competition from Massey-Harris, Claas and others the project was shelved.

With favourable reports from the Ministry of Agriculture, the trailed McCormick International No 62 combine and a batch of twenty No 123-SP self-propelled machines were sent to the UK in 1946 under the Lend-Lease scheme. Launched in 1942, the McCormick-Deering 123-SP had a 45 hp International petrol engine, a 12 ft cutter bar, a 20 in wide threshing drum and four straw walkers. Although no more 123-SP combines were sold in the UK, more than 10,000 had been made when it was replaced by the McCormick-Deering 125-SP in 1948.

Following a ten-year production run the McCormick Deering No 62 was superseded by the McCormick International No 64 in 1951. The Doncaster-built McCormick International B64 was launched at the 1952 Royal Smithfield Show and was made until 1964. In common with other trailed combines on the market the pto-driven B64, with a bagging-off platform or grain tank, had a 6 ft wide cutter bar with a slightly narrower threshing cylinder, straw rack and sieves. Optional equipment included a Petter PAV4 air-cooled engine, a weed seed screen and a pick-up attachment. Mid-1950s additions to the list of optional extras for the B64 included an Armstrong Siddeley diesel engine and hydraulic table lift.

European production of McCormick International combines had moved to France by 1960 when the F8-63 self-propelled combine was advertised as a two-in-one combine with a grain tank and a bagging-off platform. The specification included a 6 ft 6 in wide cutter bar, a 27 in wide threshing drum, three straw walkers and a 35 hp International diesel engine.

1.20 The Grain Marshall 626 combine was made at Gainsborough between 1952 and 1958.

The self-propelled McCormick International 8-41 tanker combine, introduced to UK farmers in 1964, and the slightly later 8-51, still available in the early 1970s, had an 8 or 10 ft floating cutter bar controlled hydraulically through a nitrogen accumulator. Provision was made for the tailings on the 8-41 and 8-51 to be returned either to the cylinder for re-threshing or on to the grain sieves. In difficult threshing conditions it was possible to divide the tailings equally between the drum and sieves.

Many farmers with large acreages of cereals were using a self-propelled combine harvester in the mid-1950s. A survey of combine harvesters on the UK market in 1959 listed twenty-six machines from twelve manufacturers and importers, of which nineteen were self-propelled. The others, made by Aktiv, Allis-Chalmers, Claas, David Brown Albion, International Harvester and Ransomes, were trailed machines with engine or pto drive and a 5, 6 or 7 ft wide cutter bar.

Harrison, McGregor & Guest announced a trailed 4 ft cut Albion combine harvester at the 1953 Royal Smithfield Show. The pto-driven Albion combine harvester had a ground wheel-driven bat reel, a 3 ft 1 in wide threshing drum with rubber-faced beater bars, a straw rack and a bagging-off platform. Little more was heard of the Albion combine and the project was shelved when the David Brown organisation acquired Albion in 1955.

David Brown exhibited a new 5 ft cut trailed Albion combine at that year's Smithfield Show. Designed and built by Aktiv in Sweden the pto-driven combine had a vee-belt-driven 3 ft 11 in wide drum with steel rasp bars and a full-width straw rack. Sales literature explained that as the David Brown Albion combine required less than 30 hp at the power take-off, having an expensive engine-driven combine harvester standing idle for 85 percent of the year would be a thing of the past. Optional equipment included a petrol engine, a 2 ton grain tank, a pick-up attachment, a pre-cleaner and sieves for combining small seeds.

Ransomes, unconvinced about the future of the combine harvester, met a request from a Yorkshire farmer to build a combine based on their well-tried threshing machines. A pick-up cylinder and elevator at one side of the machine lifted the previously cut crop to the self-feeder on top of the threshing drum where a man cut the bonds and fed in the sheaves. Another man standing on a platform bagged up the grain, while the chaff and straw were left on the ground. A petrol engine on top of the threshing drum provided power.

Ransomes eventually decided there might be a future in the combine harvester market and decided to import some trailed and pto-driven Bolinder Munktell combines. Ransomes imported a batch of MST42 combines from Sweden to meet existing customer requirements and then announced it would make the combine under licence for the 1954 harvest. The MST42, which required a 28 hp tractor, had a 42 in

1.21 The 5 ft cut David Brown Albion combine harvester was launched at the 1955 Smithfield Show.

wide drum with six beater bars and a 6 ft 3 in long straw rack to separate out any remaining grain in the straw. Sales literature explained that the MST42 had a spacious bagging-off platform and an optional pick-up attachment was available for harvesting previously windrowed crops. Ransomes kept faith with tradition by using the term 'drum' from the early days of threshing machines in preference to 'thrashing cylinder' and always spelt the word 'threshing' with an 'a' rather that the more usual 'e'.

The 5 ft cut MST56 replaced the MST42 in 1959 when the Agricultural Machinery Census recorded a total of 48,000 trailed and self-propelled combines and 10,500 threshing machines on farms in England and Wales. The pto-driven MST56, which required a 30 hp tractor, had the same size drum and straw rack as the earlier model. There was a choice of a bagging-off platform or a grain tank; a Wicksteed or Reynolds pick-up reel was available as an optional extra and a rotary screen to 'eliminate weed seeds' was available for the bagger model. An MST combine with a Ford Zephyr engine was built in an attempt to compete with the engine-driven B64 but the project was dropped and the last batch of MST56 combines was made in 1961.

The 12 ft cut M-H780, introduced in 1953, was a modified version of the earlier M-H726. Available with a six-cylinder Austin or four-cylinder Morris petrol engine the M-H780 had the same electric table lift and transmission as the M-H726. An improved M-H780 with hydraulically operated traction variator pulleys and hydraulic table lift for the 8 ft 6 in, 10 or 12 ft cutter bar appeared in 1956. Optional equipment included a Newage Austin petrol or tvo engine or a Perkins L4 diesel, a pick-up attachment, a straw spreader and a Raussendorf straw press.

Launched in 1956 the self-propelled 6 ft cut Massey-Harris 735 bagger combine was designed for the smaller acreage farm. It had a 24 hp four-cylinder Austin two power unit and, with an overall width of only 7 ft 7 in, the M-H735 was ideal for negotiating narrow country lanes. New farm safety regulations were coming into force at the time. The operator's manual suggested that the spare round hole sieves

1.22 The Ransomes MST56 replaced the MST42 in 1959.

should be kept in their brackets on the bagging-off platform to serve as guards for the chain and belt drives. When Massey-Harris merged with Ferguson the 735 and 780 combine harvesters were sold under the Massey-Harris name until 1958 when the 6 ft cut bagger or tanker MF735 cost £935. Prices started at £1,675 for the tvo-engined MF780 with an 8 ft 6 in cutter bar.

Massey Ferguson exhibited an improved MF780 Special and the new MF892 at the 1960 Royal Smithfield Show. A choice of an Austin Newage tvo or a Perkins diesel engine was available for the MF780

1.23 The 6 ft cut Massey-Ferguson 735 was designed for the smaller acreage farm.

1.24 The Massey Ferguson 788 with a tvo or a diesel engine was suitable for farms with up to 200 acres of cereal crops.

Special with a more efficient cutter bar and more user-friendly driving controls. The MF892 with a six-cylinder Perkins diesel engine was the largest Massey Ferguson combine yet seen in the UK. The 9 ft 6 in and 12 ft cut MF892 had a full set of hydraulic controls, a 3 ft wide threshing drum and a large number of sealed bearings.

Massey Ferguson announced the six straw walker MF500 and four straw walker MF400 combines in time for the 1962 harvest. Made at Kilmarnock the new tanker combines with Perkins engines were Massey Ferguson's answer to the challenge from the many imported machines on the British market. Automatic cutter bar height control was an optional extra. The driving and harvesting controls were all within reach of the driving seat and two pannier grain tanks gave the combines a low profile.

Launched in 1963 the re-badged tanker and bagger MF788 was an updated version of the MF780 Special. The low-profile 788 had re-grouped driving and threshing controls, a repositioned radiator intake screen and a full set of guards to comply with the new farm safety regulations.

While the factory at Harsewinkel was involved in wartime armament production, Claas design engineers were busy developing the trailed Claas Super combine. When the Super had cut and threshed the crop, the straw and remaining grain was turned through a right angle on to the straw walkers and grain sieves. After shaking out the rest of the grain the straw was tied in lightweight trusses. German farmers tested three Claas Super combines in 1946 and a small batch was made for the 1947 harvest season. The Claas Super required three operators – a tractor driver, a man to bag off the grain and a third to ride behind the cutting platform to raise and lower the cutter bar and reel.

D Lorant Ltd at Watford, which was already selling re-badged Claas pick-up balers, declined the opportunity to market the Claas Super. J Mann & Son at Saxham, which had previously sold Lanz tractors, was eventually appointed the Claas combine importer for the UK and in 1949 it sold 220 Claas combines.

The 5 ft 6 in cut Super Junior, intended for small farms, was added in 1953 followed by the Super 500 in 1955. The Super Automatic and Super Junior

1.25 Three men were required to operate the Claas Super combine harvester.

Automatic, introduced in 1958, were the first Claas combines with hydraulic rams to raise and lower the cutter bar and reel. Standard equipment for the Claas Super and Junior models included a swing ram straw trusser using binder twine.

The Hercules, launched in 1953, was the first Claas self-propelled combine harvester. However, as the Hercules name was already being used by another manufacturer it had to be rebadged – as the SF (Selbst-Fahrer or self-propelled) – to avoid a possible lawsuit. The 7 tons an hour Claas SF had an 8 ft 6 in cutter bar, a 50 in wide drum and four straw walkers. The 8 ft 6 in, 10 or 12 ft cut SF55, so called because it superseded the SF in 1955, had a new three forward and one reverse transmission and hydraulically adjusted cylinder speed.

1.26 The 56 hp SF was the first self-propelled Claas combine.

1.27 The Claas Matador was launched in 1961.

The Claas Huckepack self-propelled implement carrier launched in 1957 with a 12 hp flat-twin diesel could, with various attachments, be used as a self-propelled seed drill, cutter bar mower, crop sprayer, potato digger and combine harvester. In combine format a separate 34 hp VW petrol engine was used to drive the 7 ft cutter bar and threshing unit. An advertisement for the Huckepack explained that farmers could at last buy a combine harvester that need not stand idle for eleven months of the year. However, it failed to sell in big numbers and was discontinued in 1960.

The low-profile 7 ft cut Claas Europa and 6 ft cut Columbus bagger and tanker combines appeared in 1958. The first Europa with a Mercedes diesel engine had a 32 in wide drum and three straw walkers. Later Europas were supplied with a 45 hp Perkins diesel or a 38 hp air-cooled VW petrol engine. Optional equipment for the 1½ tons an hour Columbus included a 27 hp air-cooled VW petrol or 34 hp Perkins diesel engine and a straw trusser.

The 8 tons an hour Claas Matador with an 87 hp six-cylinder water-cooled diesel engine and the option of a 10, 12 or 14 ft wide cutter bar replaced the SF in 1961. The specification included a three forward and one reverse gearbox with a hydraulic variator vee-belt pulley drive with speeds from just under 1 mph to a top road speed of 10 mph. To meet demand for a smaller combine Claas added the 62 hp Matador Standard with shorter straw walkers and a 10 ft cutter bar in 1962. The original Matador became the Matador Gigant and from 1964 the Claas silver paintwork was changed to Austral green.

To fill the gap between the Europa and the Matador Claas added 8 ft 6 in and 10 ft cut bagger and tanker versions of the Mercury (or Mercur) in 1963. The four straw walker Mercury, with a 52 hp Perkins engine and 42 in wide drum, was advertised as a 4 tons an hour-plus combine for under £2,000.

To compete with the Massey Ferguson 415 and 515 tanker combines Claas introduced the improved 10 tons an hour Senator with 10-20 ft wide cutter bars in 1966. The Mercator with hydrostatic steering and a quick-attach table was added in 1967 and the 8 ft 6 in and 10 ft cut Consul followed in 1968. A useful cutting height indicator, attached to the hand brake lever, was a standard fitting on the Senator, Mercator and Consul.

Other late 1960s Claas combines included the Comet and Cosmos, which were updated versions of the earlier Europa and Columbus, the Corsar, the

1.28 The 6 ft cut Lanz MD180 with a 4 ft wide drum had an output of up to 12 acres in a 10-hour day.

Protector and the trailed Garant. Unlike earlier trailed Claas combines the Garant was a straight-through machine with an 8 ft cutter bar. The Consul, Mercator and Senator with blue and grey paintwork were badged as the Ford 620, 630 and 640 for the North American market. The first of a dozen or so models of the Claas Dominator combine with 10-17 ft wide cutter bars appeared in 1971 when optional equipment for Claas combines included a pick-up attachment, a straw chopper, a straw spreader and a cutter bar trailer.

The German-built Lanz 6 ft cut self-propelled MD240S was introduced in 1954 and trailed MD180 combines were imported by H Leverton Ltd at Spalding and Fred Myers Ltd at Windsor in the mid-1950s. The hydraulic cutter bar and the reel height adjustment were features of the Lanz MD240S. Sales literature for the MD180 explained that the pto-driven combine required a 35 hp tractor but a 16 hp tractor would suffice for the engine-driven model. The 7 ft cut MD255 with a Perkins engine and two-speed transmission with a top speed of 9 mph was added to the Lanz combine range in time for the 1960 harvest season.

John Deere, which made its first combines in 1927, realised that its 1950s range of American-built self-propelled combines was too big for most European farms. It overcame the problem in 1956 by acquiring a major share of Lanz and the re-badged John Deere Lanz 250 gave the American combine manufacturer a foothold in Europe. The self-propelled 6 ft 6 in cut John Deere Lanz MD18S introduced to British farmers in 1958 had a 32½ hp Mercedes diesel engine, a 16 in wide drum, five straw walkers and a suction-operated re-cleaning system. Optional equipment for the MD18S included a twin-knotter straw baler, a pick-up attachment and a feed table for stationary threshing at a stack or in the field. A joint H Leverton and Fred Myers stand at the 1960 Royal Show hosted the launch of the latest 6 ft cut Lanz MD255 combine harvester with a Perkins engine and a two-speed transmission with a top speed of 9 mph.

John Deere launched its first combines specifically designed for European conditions in 1964. Built at the former Lanz factory the self-propelled 30 Series combines had 48-95 hp engines and 8 ft 6 in-14 ft wide cutter bars. Publicity material stressed that particular attention had been given to driver comfort with one feature being sprung upholstered seats that could be folded away when the driver chose to work in a standing position.

By 1970 John Deere was making five self-propelled combines and one trailed model at the Zweibrücken factory. There was a choice of a 7 or 8 ft 6 in cutter bar

1.29 The John Deere 730 had five straw walkers and a 45 cwt capacity grain tank.

for the 48 hp 330, and a 40 hp tractor was recommended for the 7 ft 9 in cut trailed 360 combine. The later 730 with a 110 hp engine, a 14 or 18 ft cutter bar, a 52 in wide threshing drum and five straw walkers was the largest model in the company's European range.

Early 1970s John Deere combines for the European market, with 'Quick-Tach' cutter bar tables and John Deere engines, ranged from the 960 with 8 ft 6 in, 10 ft or 12 ft cutter bars to the 12-18 ft cut John Deere 970. Driving aids on the more sophisticated John Deere 970 included cutting height indicator, hydraulic table control with guidance skids and Revermatic transmission with an automatic reverse drive. The 965H, introduced in 1975, was advertised as Europe's first hillside combine harvester.

Jones Balers at Mold in North Wales entered the combine harvester market when it launched the self-propelled 8 ft cut Pilot in the mid-1950s. The 2 tons an hour Pilot with a 34 hp David Brown two engine had a 48 in wide drum, a one-piece straw rack, a bagging-off platform and dual front wheels. The 8 ft 6 in cut Jones D810 Cruiser combine, which cost £1,500, replaced the Pilot in 1958. The 3-4 tons an hour tanker or bagger diesel-engined Cruiser had the same threshing cylinder as the Pilot, four straw walkers, hydraulic variator pulleys for traction and a drum and pick-up reel. Jones combines were discontinued in 1961 when Allis-Chalmers acquired the company.

After deciding against building the Swedish-designed self-propelled Bolinder Munktell combine under licence at Ipswich, Ransomes introduced a prototype self-propelled 902 at the 1956 Royal Smithfield Show. Following exhaustive field testing Ransomes launched the 6 tons an hour 902 at the 1958 Royal Show. The specification for the 902 bagger or tanker combine included a 10 or 12 ft cutter bar, a Reynolds six bat pick-up reel, a 39 in wide drum and four straw walkers. A 62 hp Thames Trader diesel provided the power and a three forward and one reverse gearbox with a hydraulic variator drive gave the combine a top forward speed of 11 mph. A 16 ft cut export model of the 902 for countries with lighter yields than those in the UK was added in the early 1960s.

Introduced in 1963, the 8 ft cut self-propelled Ransomes 801 with an 8 ft 6 in transport width was advertised as small enough to be driven along narrow country lanes inaccessible to other combines in the same class. The four straw walker 801 with a 42 hp Perkins diesel and a 36 in wide drum incorporated the new Handi-Matic control system used to vary drum,

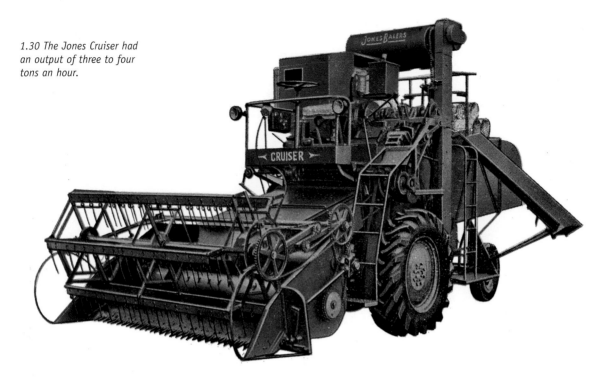

1.30 The Jones Cruiser had an output of three to four tons an hour.

reel and ground speeds from the driving seat. The 6 tons an hour Ransomes 1001 with a 10 or 12 ft cutter bar replaced the 902 in 1964. The 1001 was a scaled-up version of the 801 with the option of a four- or six-cylinder Ford diesel engine and a range of forward speeds from 0.8 to 11 mph and disc brakes for improved stopping power.

A prototype of the revolutionary twin-drum Ransomes Cavalier was exhibited at the 1965 Royal Smithfield Show and the first production machines were built in time for the 1967 harvest. Ransomes advertised the 9 tons an hour Cavalier with its 'Early Action' twin drum and concave threshing system as a new concept in combine harvester design. The 8¾ in diameter front drum was claimed to thresh out up to 25 per cent of the grain from the straw before it reached the main 24 in diameter drum and concave. There was a choice of a 12 or 14 ft cutter bar and it had a chassis-mounted 90 hp six-cylinder Ford diesel engine. It was also the first combine harvester with hydrostatic steering and one of the first to have an optional all-weather driving cab.

The Ransomes Crusader, a smaller version of the Cavalier with a 74 hp Perkins engine, twin-drum threshing and a 10 or 12 ft cutter bar was announced at the 1967 Royal Smithfield Show. The Super Cavalier, with the option of a 10, 12 or 14 ft quick-release table, superseded the Cavalier in 1974. The new model had a 102 hp Perkins diesel engine, 45 in wide twin threshing drums, four straw walkers and a 2 ton grain tank. When Ransomes were making the last Super Cavalier combines in 1976 they were also field testing a batch of prototype 16 ft cut Saracen combines based on the Cavalier and Crusader design but with wider twin threshing drums. However, as funding to develop the new model and support for the only British-owned combine harvester manufacturer was not forthcoming the project was shelved. The remaining stock of chassis and other suitable components was used for Ransomes Hunter self-propelled sugar beet harvesters.

Allis-Chalmers, which made the first British-built trailed All-Crop 60 combines at Essendine in 1951, bought the American combine manufacturer Gleaner Baldwin in 1955. Twenty self-propelled Gleaner Model A combines were imported and sold in 1958. The Gleaner Model EA tanker combine, the British version of the Model A, was made at Essendine between 1958 and 1962. With the choice of an 8 ft 6 in, 10 ft or 12 ft cutter bar the EA could have an Austin Newage or Perkins P6 engine. Unlike other combines with an

1.31 The Ransomes Cavalier was the first combine harvester with a twin-drum and concave threshing system.

elevator carrying the crop up to the threshing drum, the Gleaner drum was immediately behind the cutter bar. There were hydraulic controls for the cutter bar, reel and transmission variator pulleys. Different sprockets were used to change the speed of the threshing cylinder.

Two new models of Allis Gleaner combine, introduced in 1962, had silver paintwork and low-level threshing cylinders. The Model A Super Gleaner tanker combine with a 70 hp Perkins diesel or Newage/BMC petrol or two power unit had a quick-detach cutter bar and a 30 in wide threshing cylinder. An 84 hp Perkins engine was used for the Model C with a 1½ ton grain tank and a quick-detach 10 or 12 ft cutter bar. Sales literature described the higher output Model A Super Gleaner, designed to work under the most rugged conditions, as the most versatile harvester ever offered to farmers.

The Gleaner Super A and Super C, which superseded the earlier models, were made from 1964 to 1968. Designed for 'speedy harvesting' both combines had 40 in wide threshing drums, full hydraulic controls and floating 10, 12 or 14 ft cutter bars controlled by a hydro-pneumatic accumulator. The Gleaner CH Hillside Special had an extra-wide wheel track and the cutter bar worked at an angle on sloping ground. A hydraulic suspension system kept the threshing and separation systems in a horizontal position. According to sales literature the CH Hillside Special was the biggest capacity combine on the market with its sure feed system, down-front cylinder and exclusive two-fan cleaning system.

The Gleaner Super A was still made in 1968 when Allis-Chalmers launched the Model 5000 combine harvester. The Model 5000 followed the Gleaner tradition of having a 40 in wide 'down-front' threshing drum behind the cutter bar. Other features included a 103 hp diesel engine, four straw walkers, hydrostatic steering and hydraulic disc brakes. Bamfords bought the Allis-Chalmers factory at Essendine in 1971 and within a few months discontinued the Model 5000.

The mid-1950s trailed 5 ft 6 in cut Swedish Aktiv Model M for 25 hp tractors was imported by Western Machinery & Equipment of Ivybridge in Devon. The straight-through Model M with a 4 ft wide threshing cylinder and straw rack was available with a bagging-off platform or a 1 ton grain tank. Described in sales literature as the only combine in the world with fingertip-controlled cylinder speed, the Model M had outputs of up to 3 tons an hour.

1.32 A quick-detach cutter bar was a feature of the Allis Gleaner Model C combine harvester.

The early 1960s self-propelled Aktiv M2S advertised as a double-rub harvester with two 4 ft wide threshing drums and concaves had a 7 ft 8 in cutter bar, a straw rack and a 42 bushel grain tank or a bagging-off platform. After being threshed by the main drum and concave the crop passed to a second unit where any remaining grains were removed from the straw. The first Aktiv M2S combines had a 35 hp Mercedes-Benz diesel engine, while later machines were powered by a 38 hp Perkins diesel. A straw trusser, straw chopper and a windrow pick-up attachment were listed as optional extras for the M2S.

Western Machinery & Equipment introduced the self-propelled 8 ft cut Aktiv 800 combine in 1965. The 4 tons an hour Aktiv 800 had a 40 hp Perkins engine and a 40 bushel grain tank. An advertisement explained that daily maintenance had been reduced to a single item – oiling the knife! The self-propelled Aktiv 1000 introduced in 1967 with a 9 ft cutter bar, a 41 in wide drum and a 58 hp Perkins engine was the biggest model in the early 1970s range of Aktiv combines. The Swedish plough maker Överum sold a range of 60-84 hp Aktiv combine harvesters to UK farmers in the early 1980s.

The first Swedish Viking ST67 self-propelled bagger and tanker combines were imported in the mid-1950s by Farm Machinery & Accessories Ltd of Ashburton in

1.33 The Model 5000 was the last Allis-Chalmers combine built at Essendine.

1.34 The Aktiv Model M with working speed of up to 5 mph harvested up to 2½ acres an hour.

Devon. The 7 ft cut ST67's huge 5 ft 11 in wide threshing drum, sieves and straw rack were driven by a 52 hp Ford diesel engine. Viking Farm Machinery at Ivybridge in Devon, which later became part of the BM-Volvo group, exhibited the self-propelled ST256, ST68 and ST610 combines with 6 ft 3 in, 8 ft and 10 ft cutter bars at the 1960 Royal Smithfield Show.

Viking Farm Machinery was selling a range of Viking Bolinder-Munktell tanker and bagger combines, including the 10 and 12 ft cut S1000 and the ST257 with 7 and 8 ft cutter bars, in 1962. The 9 tons an hour S1000 with a 79 hp Volvo direct injection diesel engine and a full range of hydraulic controls had a 43 in wide drum and four straw walkers. The S1000 and the smaller ST257 with a 35 hp Perkins engine, a 30 in wide drum and three straw walkers had hydraulic table lift and variable belt drive transmission.

Lesser-known combine harvesters in the 1960s included those made by Dania, Fahr, Fisher Humphries, JF, Koela and Lely. Trailed and self-propelled Dania combines made by Dronningborg in Denmark were imported by Shermore at Norwich. They ranged from 5 and 6 ft cut trailed models to the self-propelled SP41, SP50 and the Great Dane. The 9 tons an hour Great Dane with a 10 ft cutter bar, a 44 in wide threshing drum and four straw walkers was the first combine with a built-in monitoring system which warned the driver of overloaded grain sieves or grain elevator blockages.

Western Machinery Ltd at Ivybridge imported a range of Dania combines in the early 1970s, including the self-propelled Dania D2500 with a 135 hp Perkins diesel, hydrostatic drive for the reel and knife and hydraulically pre-set cutting height on the optional 16, 18 or 20 ft cutter bar. Early 1970s prices for the D2500 started at £5,850 but within four years the cost of the combine with a 150 hp engine had more than doubled to £13,500.

The first Fahr combines were made in Germany in 1939. The Fahr MDT was introduced in 1955 and Fahr Products Ltd at Ivybridge in Devon imported the self-

1.35 The Viking ST257 combine was imported by Viking Farm Machinery from Sweden in the early 1960s.

propelled Fahr MD4, MDL and MD5A combines in the late 1950s. The Fahr MD4 had a 59 in wide drum with a similar width straw rack and grain sieves.

The four straw walker Fahr M66 and M88 bagger and tanker combines with Deutz air-cooled diesel engines cost £2,325 and £2,685 respectively in 1965. The 7 ft 6 in and 8 ft 6 in cut M66 had a 52 hp engine and a 2 ft 10 in drum. The Fahr M88 with a 9 or 10 ft cutter bar and a 3 ft 5 in wide drum had a 1½ ton grain tank.

Deutz acquired Fahr in 1968 and by the early 1970s a range of Deutz Fahr combines, including the M1200, was imported by Watveare Overseas. The five straw walker M1200 with four cutter bar widths from 10 to 16 ft had a six-cylinder 127 hp Deutz diesel engine. The 203 hp M1600H with hydrostatic transmission and cutting widths of up to 19 ft was the biggest and most expensive in the five-model Deutz Fahr range of the mid-1970s.

Fisher Humphries at Wootton Bassett in Wiltshire, well known for its balers and ploughs, introduced the Fisher Humphries Victory combine harvester in 1965. The 14 ft and optional 18 ft cut Victory with a 53 in wide threshing cylinder, a 4½ ton grain tank and an output of up to 17 tons an hour was advertised as the world's largest combine harvester. Hydraulic rams were used to fold the two-section cutter bar table vertically in ten seconds to reduce the transport width to less than 10 ft.

Fisher Humphries became part of British Lely shortly after the launch of the Victory combine. The Fisher Humphries factory at Wootton Bassett was closed in the late 1960s and production of British Lely and Fisher Humphries machinery was transferred to Lely UK at St Neots. An improved Mk2 Lely Victory, which appeared in the early 1970s, had a larger 100 bushel grain tank. New patent Lely variator pulley drives were used for the traction and threshing drum

1.36 Made at Dronningborg in the early 1960s the 8 ft cut Dania combine had a 44 in wide threshing cylinder and similar width straw rack.

1.37 Optional equipment for the 7 ft 4 in cut Fahr MD 4A included a baler, an 18 cwt grain tank and a pick-up cylinder.

and the time taken to fold the cutter bar table was reduced to eight seconds. The mid-1970s Mk3 Victory with hydrostatic transmission remained in limited production for another ten years.

Claas had built a combine round a Lanz Bulldog tractor in 1930 with limited success and a similar prototype Ferguson combine harvester appeared in 1956. JF wrap-around combine harvesters, introduced by JF Farm Machines in Denmark in 1961, were made for the next twenty-five years. Like the first wrap-around Claas combine the JF MS5 had a 5 ft front cutter bar with the threshing and separating units mounted on the side of the tractor and a bagging-off platform at the rear. Sales literature explained that the pto-driven MS5 could be attached to a tractor in less than five minutes. The tractor was reversed into the combine and after connecting the drawbar, the combine unit was swung across on its transport wheels and secured to a frame at the front of the tractor.

Early 1970s JF side-mounted combines included the 7 or 8 ft cut MS90 and the 10 tons an hour MS105 Grainflash with a 10 ft cutter bar, hydraulic controls and a grain tank or bagging-off platform. More than 25,000 4-10 ft cut tractor-mounted JF combines had been built when the last one was made in 1986.

1.38 The Fahr M88 combine designed for medium-sized farms had a 64 hp Deutz diesel engine.

Colman & Co (Agricultural) in London imported the German Kola self-propelled combines, badged as Koela in the UK, from the early 1950s. Made by Ködel & Böhm the self-propelled 6 ft 3 in cut Koela Combi bagger with a 34 hp Mercedes diesel engine cost £1,590 in 1959. Optional extras for the 3 tons an hour Koela Combi included hydraulic table lift, a grain tank and a baler attachment.

Three Koela combines including the 6 ft 3 in cut Combi Standard with a bagger platform and bagger and tanker versions of the Koela Favourite were current in the mid-1960s. The Koela Hydromat introduced in the late 1960s had a 110 hp diesel engine, a 10, 12, 14 or 16 ft cutter bar, a 50 in wide drum and an 80 bushel grain tank. Sales literature stated that the Koela Hydromat was the world's first high-capacity combine with a fully hydrostatic transmission.

Leon Claeys was making trailed combine harvesters at Zedelgem in Belgium in 1949 and the Claeys MZ100 launched in 1952 was the first European-built self-propelled combine harvester. AB Blanch at Crudwell introduced the Claeys MZ combine to British farmers in 1955. Bamfords entered the combine harvester market in 1958 when it became the UK distributor for the Claeys MZ100 and the new self-propelled Claeys M73 combine harvester. The 10 and 12 ft cut MZ100 and the M73 with a 7 ft 3 in cutter bar were five straw walker machines with a 24 in wide threshing drum. The Bamford Claeys M103, with an 80 hp diesel engine, hydraulically raised cutter bar and reel and double disc brakes, made its debut at the 1959 Royal Smithfield Show. With 8 ft 6 in, 12 ft and 14 ft wide cutter bars prices for the M103 started at £2,275.

1.39 An optional 135 hp turbocharged Perkins diesel was available for the MkII Lely Victory combine harvester.

Bamfords had sold more than 3,000 Claeys combines ranging from the 8 ft cut M80 to the 19 ft cut M140 by 1964 when Sperry New Holland bought the Claeys combine factory. While Bamfords provided field support for the British fleet of Claeys machines during the 1964 harvest it also acquired Viking Farm Machinery which held the UK franchise for BM Volvo combine harvesters.

Bamfords exhibited the Bamford BM Volvo ST257 and the improved S1000A along with the Bamford Landlord 55 at the 1964 Royal Smithfield Show. Laverda, which made its first self-propelled combines in 1956, produced the Landlord 55 in Italy. The 9 tons an hour Landlord 55 with a 105 hp Perkins engine, a 50 in wide drum, five straw walkers and a £3,720 price tag was one of the most expensive combines on the market. An advertisement pointed out that the Landlord 55 had eleven grease nipples compared with eighty on the Massey Harris 780 – and only one of them required daily attention with a grease gun!

Bamfords introduced the 117 hp BM Volvo S950 with a six-cylinder Volvo diesel engine and the choice

1.40 JF Farm machines introduced the side-mounted MS5 combine in 1961.

1.41 The Claeys MZ100 was the first European-built self-propelled combine harvester.

of a 12, 14 or 16 ft cutter bar in 1966. The smaller BM Volvo S900 with a 78 hp Perkins engine appeared in time for the 1967 harvest season. Features of both combines included a 'comfortable driving seat with good all-round visibility', a quick-release cutter bar table with hydro-pneumatic suspension and a mechanical variator pulley in the belt drive to the pick-up reel.

Following the addition of the smaller BM Volvo S830 with a 56 hp Perkins power unit in the late 1960s Bamfords acquired the British arm of Allis-Chalmers in 1971. Within a few months the Allis-Chalmers 5000 combine was discontinued in favour of its existing range of Bamford BM-Volvo machines. When Bamfords lost the BM-Volvo combine franchise in 1974 the company restored the Laverda franchise. Bamford Laverda models current in the mid-1970s ranged from the compact 94 hp M84 with 7 ft 3 in-10 ft cutting widths to the M150 with a 14, 16 or 18 ft wide cutter bar. Bamfords imported six models of Laverda combine in the late 1970s. The largest, the M152 with a 14 or 16 ft cutter bar, had a 130 hp diesel engine, a 53 in wide threshing drum, five straw walkers and a large capacity grain tank. Bamfords sold Laverda combines in the UK until 1981 when a Syrian businessman bought Bamfords and the Uttoxeter company became Bamfords International.

Most mid-1970s combine harvesters, including those made by Bamfords, Claas, Dania, Fahr, International Harvester, John Deere, Massey Ferguson and New Holland, were tanker models. A cab, a quick-release cutter bar table, hydrostatic steering and an in-cab performance monitoring system were standard on the larger combines, and hydrostatic transmission was either standard or optional on the more expensive machines.

The International 321, 431 and 531 tanker combines with optional cabs replaced the 8-41 and 8-51 in the mid-1970s. The three new models with 8-14 ft wide cutter bars were made in France. The late 1970s International 953 with a 140 hp turbocharged diesel engine had a 14 ft cut quick-detach table, a 52 in wide drum and five straw walkers. An aspirator dust mask and a cab were among the list of extras and later options included a 150 hp engine and hydrostatic transmission.

The 1967 Massey Ferguson catalogue included the MF410, 415, 510 and 515 combine harvesters that were

1.42 The Bamford BM-Volvo S950 combine had a 117 hp Volvo diesel engine.

updated versions of the earlier MF400 and 500, the 7 ft cut MF31 and the ageing MF788. There was a choice of a grain tank or bagging-off platform for the MF31 and an optional extension tube for the tank unloading auger spout enabled an operator to bag up the grain while standing on a trailer. A larger straw walker area and optional automatic table height control were among the improvements on the new 410 and 510.

The 10 tons an hour MF515 Multi-Flow and smaller 415 Multi-Flow had hydraulic accumulators instead of the usual ram springs for improved cutter bar flotation. Multi-Flow was a double separation system which, instead of dropping the straw over the back of the straw walkers, directed it down a sloping pan to a rotary beater where any remaining grains were shaken from the straw before it was returned to the stubble. The

1.43 International Harvester 431 combine harvesters were made in France.

new combines, like the earlier 400 and 500, had saddle grain tanks to reduce their overall height and improve stability.

The Massey Ferguson 525 and 625 Multi-Flow and the smaller 187 and 487 combines were launched at the 1970 Royal Smithfield Show. An optional telescopic cutter bar table system with two easy-to-remove central sections was available for the MF187 and MF487. A crank handle mechanism was used to adjust the reel and the auger to match the width of the cutter bar. As well as varying the cutting width to suit the condition of the crop the overall transport width could be reduced without removing the table when driving along narrow lanes or through field gateways.

The MF525 and 625 Multi-Flow combines with 10-16 ft wide quick-attach tables were still being made in 1975 along with the new MF307 for the smaller farm, the MF506 and the high-capacity MF760. Sales literature explained that the 77 hp MF307 had plenty of power to spare for hillside work and to drive a straw chopper. Unlike other MF combines the six straw walker 760 had a manual or optional hydrostatic transmission and a de luxe driving cab with full instrumentation including an electronic speed sensor for the 60 in wide threshing cylinder.

The Claas Senator was renamed the Mercator 70 in 1968 and five years later the 8 ft 6 in cut and the 10 ft cut Claas Protectors became the Mercator 50 and Mercator 60 respectively. The Claas Compact with 7 or 8 ft cutter bar for small acreage farms appeared in 1970, as did, at the other end of the scale, the 15 ft cut Claas Dominator 80. Various sizes of Dominator combine were made for the larger acreage farm throughout the 1970s and into the early 1980s with the Consul, Compact and Mercator meeting the needs of small and medium farms.

Bonhill and Ursus Bizon imported limited numbers of Eastern European combines in the 1970s and 1980s. Bonhill Engineering at Thetford in Norfolk distributed the E516 and smaller E514 Fortschritt combines from East Germany. The Fortschritt E516 had a 228 hp V8 diesel engine, a 66 in wide drum, a 2½ ton grain tank and the choice of a 22 or 25 ft wide cutter bar. Ursus Bizon combines with Leyland diesel engines included the 118 hp Z-056 with a 10, 14 or 17 ft cutter bar and the 17 or 20 ft cut 220 hp Z-060 with a 66 in wide threshing cylinder.

1.44 The Massey Ferguson 500 had outputs of up to 7½ tons an hour.

1.45 The Massey Ferguson 487 had a work rate of up to 4 acres in an hour.

The Claeys M80, M103 and M140 with Ford diesel engines were current when Sperry New Holland bought the Zedelgem factory in 1965. The M140 Armada with a 13-19 ft wide cutter bar was a five straw walker machine with a 50 in wide drum. Five New Holland Clayson combines from the smallest 8 ft 6 in and 10 ft cut Clayson 1220 to the 17 ft cut Clayson 1550 were current in the early 1970s. Optional equipment for the 125 hp 1550 included a grain tank or bagging-off platform, a peg drum option for North American farmers, tracks, rice tyres and a six-row maize header. Clayson combines were popular in America and about 7,000 Clayson 975, 985 and 995 combines with red and yellow paintwork were sold there between 1965 and 1974.

1.46 A sunshade was an optional extra for the New Holland Clayson 140 Armada combine harvester.

1.47 An air-conditioned cab was available for New Holland Clayson 8070 and 8080 combine harvesters.

The 1973 New Holland Clayson S1550 had a rotary separator behind the threshing cylinder to remove more grain from the straw before it reached the straw walkers. The 8000 series of Clayson combines first appeared in 1977. The top of the range 15-22 ft cut 8080 had a 62 in wide drum with a rotary separator, six straw walkers and a 170 hp Mercedes V6 diesel engine.

Into the 1980s

Combines with rasp bar cylinder and concave threshing, straw walkers and round hole or adjustable sieves were still in fashion in the 1980s when some had an output in excess of 18 tons an hour. Engines had passed the 200 hp mark, electro-hydraulic controls were replacing levers and knobs, some of the more expensive combines had a hydrostatic transmission and the rotary combine had arrived on the UK farming scene.

Having spent a fortune developing and testing a rotary threshing system, employing a principle that dates back for the best part of 100 years, International Harvester introduced the Axial Flow combine to American farmers in 1978. Instead of threshing the crop with a rasp bar cylinder, straw walkers and grain sieves the Axial Flow had a longitudinal rotor with spiral rasp bars inside a 360 degree concave that threshed the crop against a full-length concave. The threshed grain dropped through the concave grate on to the grain sieves and the straw was returned from the end of the rotor on to the stubble.

A few Axial Flow combines were field tested in the UK, France and Germany in 1979 and the International Harvester 1400 series was awarded a gold medal at that year's SIMA Show. The new rotary combines were such a success in North America that by 1981 International Harvester combine production in America was restricted to Axial Flow machines. American farmers had the choice of four models of International Harvester 1400 Series Axial Flow

1.48 The International 1480 Axial Flow combine had a 210 hp six-cylinder diesel engine.

combine in 1982. The smallest 1420 with 10-20 ft widths of cut had a 112 hp engine and the top of the range 1480 with a 210 hp power unit had a 13-30 ft wide cutter bar. A new International Harvester factory in France was opened in 1983 to build Axial Flow combines for the European market. Three models of the 1600 series, the second generation of Case IH Axial Flow combines, with 15-20 ft wide cutter bars, superseded the 1400 Series in 1986.

Most combine harvester manufacturers were making at least one model with an unconventional threshing or separating system in the early 1980s. The 20 ft cut Claas Dominator 116CS with a rasp bar threshing cylinder and concave, launched in 1981, followed the trend. A set of eight rotary separating cylinders and concave grates under the rear hood replaced the straw walkers. Grain shaken from the straw fell through the concave grates on to the conventional grain sieves and fan separation system. The 112CS and 114CS with hydrostatic transmission were added in 1983 and the later 115CS, with 52 in wide threshing drums, had a central information system in the cab to monitor the various combining functions. The 116CS grain tank held 6½ tons before the driver in his air-conditioned cab needed to call up a grain trailer, probably on his CB radio.

More recent Claas combines include the Commandor CS range, introduced in 1986, with 15-20 ft wide cutter bars and eight straw separating cylinders. Dominator Mega combines launched in 1994 and the 1997 Lexion range both had the Claas APS (Acceleration and Pre-separation) system with twin threshing cylinders and concaves. The Lexion also had twin longitudinal rotors instead of the straw walkers to shake out any remaining grain in the straw.

Three models of the New Holland TF combine launched in 1983 had Twin Flow separating rotors behind the threshing cylinder instead of straw walkers. The crop was divided into two as it passed into the twin flow rotors where any remaining grain was shaken from the straw as it made its way to the back of the combine.

A self-levelling cleaning shoe and larger grain tank were features of the New Holland TX range. Introduced in 1986 the five and six straw walker TX combines automatically maintained the pre-set cutting height and kept the cutter bar parallel with the ground.

The Deutz M2780 with a mechanical transmission and the hydrostatic M2780H were the biggest Deutz Fahr combines in the early 1980s. Cutter bar widths from 10 to 16 ft were available for both versions of the

1.49 The Dominator Mega with APS was one of five new Claas combines launched in 1994.

five straw walker combine with a 160 hp air-cooled Deutz diesel engine and a 52 in wide threshing drum. The Agrotronic system in the cab monitored the various combining functions and warned the driver if anything went wrong.

The North American Massey Ferguson 8590 rotary combine with a 300 hp Perkins V8 engine, a 12 ft long threshing rotor and an 18, 20 or 22 ft wide cutter bar was demonstrated to British farmers in 1987. The MF8590 was very expensive and with farm finances in the doldrums the big American combine was soon withdrawn from the UK market.

Ten Massey Ferguson combines, ranging from the 5 ft cut MF8 plot combine to the Canadian-built 184 hp MF865 with a 16 or 18 ft cutter bar and six straw walkers, were current in the early 1980s. Other MF models, launched in 1984, included the 24, 27, 29 and 31 with a cutting height gauge, air-conditioned cab and hydrostatic transmission.

Launched in 1988 the MF38 made at Dronningborg in Denmark used GPS, with a computerised performance monitoring system linked to global positioning satellites in space. The computer constantly recorded the position of the combine in the field and the flow of grain into the tank. Back at the farm office the disc could be used to produce yield maps of every field on the farm.

The 4, 5 and 6 straw walker John Deere 1000 series combines, launched in 1981, had air-conditioned cabs. The biggest 195 hp 1085 with hydrostatic transmission had a 12-20 ft cutter bar and a 61 in wide drum. The John Deere 1085 grain tank held 4½ tons of wheat. The John Deere 1100 series was launched at the 1987 Royal Smithfield Show and like most high-specification combines the 180 hp John Deere 1177 had hydrostatic transmission.

Top of the range combine harvesters in the 1990s bristled with electronic gadgetry. The Ford New Holland TX34 had ultrasonic sensors to maintain the selected cutting height and an electro-hydraulic unit automatically kept the cutter bar parallel with the ground. The biggest combines, all with hydrostatic transmission, had cutter bars up to 30 ft wide, engines developing 375 hp or more and either a rotary separator or as many as eight straw walkers. Most combines had a straw chopper, and optional

1.50 The 300 hp six straw walker New Holland TX68 had a 5 ft wide threshing cylinder.

1.51 The complete operator's manual was stored in the Massey Ferguson 38 combine's in-cab computer.

1.52 The five and six straw walker John Deere Z series combines launched in 1992 had a secondary threshing cylinder.

1.53 A mid-1990s combine with GPS yield mapping system in the cab.

extras included special headers for harvesting grain maize and rubber tracks to reduce the soil compaction caused by the very wide pneumatic-tyred driving wheels.

Push button control systems made it possible to make precise adjustments to the threshing and separation systems while seated in the cab. By the late 1990s laser beams provided automatic guidance systems to steer the combine along the edge of a standing crop so that the full cutting width was always in use.

Yields averaged little more than a ton an acre in the mid-1930s when, on a good day, four men with a Case Model Q combine harvested 25 acres of wheat. Sixty years later, with yields in excess of 3 tons an acre, it took under three hours for one combine supported by a tractor and trailer to harvest the same area of wheat.

Pea and Bean Harvesters

The pea crop grown either as dried peas or vining peas for quick-freeze was harvested with a cutter bar mower. Dried peas were usually stacked on tripods to dry and either stacked or threshed off the tripods in the field. By 1950 some farmers within a relatively short distance of a pea-vining station were cutting the green pea crop with a mower and collecting the peas with a buck rake or a green crop loader. The haulm complete with pods was taken by lorry to the nearest stationary vining station where the peas were shelled and taken to the freezer factory within two hours of being cut.

Several makes of pea swather with a reciprocating knife cutter bar and a side conveyor for cutting the green pea crop came into use in the early 1950s. The crop was usually collected with a green crop loader. Most makes of pea swather, including the Hume and McBain, were pto driven and carried on the three-point linkage. As the tractor had to be driven in reverse a rear-facing seat and suitably positioned controls were supplied with rear-mounted swathers. Front-mounted pea swathers, including the JF from Denmark, cut the crop and left it in a windrow alongside the machine.

The late 1960s Hume pea swather with an 88 in wide cutter bar had a pick-up bat reel made from English oak, a fully floating cutter bar and a

1.54 *The 6 ft 6 in cut McBain pea cutter-rower was supplied with the necessary controls to drive the tractor backwards.*

1.55 A Hume pea swather followed by a Hume green crop collector speeded up the vining pea harvest.

waterproof side conveyor. It was said to cut up to 16 acres in a day. The windrowed crop could be loaded on to a trailer or lorry with a Hume green crop loader.

Stationary Mather & Platt pea vining machines with outputs of up to 30 cwt of peas an hour which were portable enough to move from farm to farm came into use in the mid-1950s. By the early 1960s the first mobile viners had arrived in the pea fields. The crop was cut and windrowed with a swather ready for the mobile viner with a pick-up cylinder and a single beater threshing chamber to collect and shell the peas. The viner, mounted on a trailer with large pneumatic-tyred wheels, was similar to the earlier static machine. Suitable for one-man operation with a Fordson Super Major or similar tractor, the viner had a two-way levelling system and a turning circle of

1.56 Self-propelled swather windrowers were in use in the mid-1960s.

1.57 Cutting and shelling peas with a mobile viner was a one-man operation.

less than its own length. Mobile viners were moved round a group of pea-growing farms making it possible to harvest the peas in the field. Transport costs were reduced as only the shelled peas had to be taken to the freezer factory and there was no stripped haulm to cart back to the farm.

By the mid-1970s self-propelled pea harvesters with a picking reel used to gather the pods and some leaves and a multi-beater threshing chamber did the complete job in a single pass. Early self-propelled pea harvesters had a diesel engine of at least 150 hp and hydrostatic transmission. By the early 1980s the FMC679 mobile viner with a 213 hp engine was harvesting about 2,000 acres of vining peas in a five-

1.58 The FMC Model 879 pea harvester had a 10 ft 6 in wide picking reel.

1.59 The Mather & Platt ST5 green bean harvester was made in the late 1960s.

week harvest season. The even bigger FMC879 pea picker, introduced in 1985 with a longer threshing cylinder than the previous model had an eight cylinder 295 hp Deutz air-cooled diesel engine. The 879 picker combed the pods off the plants with its 3.2 m wide picking reel, threshed out the peas and elevated them in a dump-on-the-go hopper.

Harvesting green beans for quick freeze or canning was done for many years with a pea cutter and swather and the cut crop was taken to a nearby processing factory but by the late 1960s trailed green harvesters were picking the beans in the field. The Mather & Platt ST5 green bean harvesters picked dwarf beans, one row at a time, cleaned the beans and put them into sacks or bulk boxes. The alternative SR6 bean harvester elevated the beans into a tipping hopper mounted above the tractor. Hydraulic motors using oil supplied from an external reservoir by a pto-mounted pump operated the picking reel and a cleaning unit. Lifters were used to guide the plants to the spring-tined picking reel which combed the bean pods off the plants. A suction fan removed the leaves and other green material as the beans were conveyed to a holding hopper.

Self-propelled green bean harvesters like the FMC GB2700 with a front picking reel harvested up to seven rows at a time. A bank of fans under the front hood removed the loose leaves and light trash before the beans were conveyed to a dump hopper behind the cab. The harvester had a six-cylinder air-cooled engine, hydrostatic transmission, an 8 ft 10 in wide picking reel and the tipping dump hopper held up to 2½ tons of beans. A four-wheel drive option for the mid-1980s FMC green bean harvesters with a turbocharged 160 hp engine and a 2.7 m working width had a work rate in excess of an acre an hour.

1.60 The dump hopper on the self-propelled FMC green bean harvester held 2½ tons of bean pods.

Chapter 2
Balers

Over the past seventy years balers have developed from stationary machines for baling hay or working behind a threshing machine to big, round and square balers that make heavy bales up to a ton in weight. A few pick-up balers, mainly of American origin, were in use in the UK in the late 1940s but on most farms a stationary high-density baler, often owned by an agricultural contractor, was used to make wire-tied bales of hay and straw.

Stationary Balers and Trussers

In the late 1800s some farmers used a hand-operated press to make bundles of hay or straw that were tied by hand with twine. By the end of the century some farmers and threshing contractors used a straw trusser attached to and driven from a threshing machine. An 1886 Ransomes catalogue explained that trusses weighing up to 36 lb could be direct loaded from the trusser on to a wagon. Ruston & Hornsby at Lincoln also made low-density straw trussers and, following their link with Ransomes in 1919, Hornsby trussers were included in Ransomes farm equipment catalogues. Horse-drawn Hornsby portable straw trussers used with a threshing machine were still being made in 1951.

Low-density trussers made bundles of hay or straw weighing no more than 40 lb. A swinging ram packed the hay or straw into bundles and instead of cutting each charge with a knife on the ram it was folded by the ram and packed sideways to form the truss. Each truss was automatically tied with binder twine.

Howards of Bedford, a member of the ill-fated Agricultural and General Engineers Association, made Howard Lion and Howard Junior hay and straw presses in the early 1920s. The Lion, belt-driven from a threshing machine, made rectangular wire-tied trusses weighing up to 20 lb. Claas was making straw trusser attachments for threshing machines in the 1920s and had sold about 15,000 by 1931 when the company introduced a stand-alone trusser driven by a belt from a threshing machine.

Ransomes, Dening of Chard, Fisher Humphries,

2.1 The twin-knotter Hornsby No 4 trusser for use with any make of threshing machine and supplied with high wheels and horse shafts cost £129 10s in 1951.

2.2 The Howard Lion made 16 x 18 x 36 in long wire-tied bales.

Jones and Wallace were among the firms making high-density stationary balers in the 1940s and 1950s. Straw was usually baled directly from the threshing machine while hay crops were collected with a hay sweep or wagon and forked by hand into the bale chamber.

The Ransomes stationary baler, described in sales literature as a high-density hay and straw baling press, was typical of the type. Straw from the threshing drum fell into the feed hopper where a set of reciprocating tines projecting through the floor carried the crop into the bale chamber. A horse's-head action packer or feeder arm pushed the straw into the bale chamber where a heavy cast-iron ram running on rails formed the bale. Boards were used to separate each bale and

2.3 The Ransomes high-density baler had an output of 15–20 tons in a day.

large twine needles were pushed by hand between the boards before wires were threaded through the needles and around the bale. The ends of the two wires were then twisted together by hand.

The Fisher Humphries high-density baler made wire-tied bales weighing up to 1½ cwt. The baler could be belt driven from a threshing drum or tractor belt pulley or with a stationary engine when hay was forked into the baler by hand. Introduced in 1946, an optional rear elevator unit could be used to load hay into the bale chamber. A hand clutch was used to disengage the drive to the elevator while the bale was secured with two bands of wire.

The Dening Somerset hay and straw baler made bales 17 in wide and 22 in high at rates of up to 2½ tons an hour. The instruction book for the Dening baler suggested that after it had been towed a long distance all the bolts should be checked as some might have worked loose through vibration from the road. The wire-tying Jones Tiger and Cub stationary balers, built at Mold in Wales and belt driven from a threshing drum, made bales 17 in high, 22 in wide and 36 in long for easy stacking. John Wallace & Co at Glasgow launched the Potts self-feed baler, which made similar size bales, at the 1948 Royal Show.

Wire-tied bales were not particularly popular with dairy farmers as short ends of wire mixed in with hay or straw could have serious consequences. Aware of the problem, and with baling wire in short supply, Fisher Humphries introduced the string-tying MkII version of its stationary baler with a two-ball twine can in 1948. Needles were used to thread twine round the bale and after cutting both twines the operator tied the ends, leaving enough slack for the bale to expand as it left the bale chamber.

The Jones Panther stationary baler introduced in the late 1940s was one of the last new stationary balers

2.4 The optional hay elevator for Fisher Humphries balers was folded away for transport.

made for the British market. Designed for use with a threshing drum, the Panther had a pair of knotters that automatically tied each bale with heavy-duty twine. Grooves, which were formed around the bale as the crop was compressed in the bale chamber, provided a recess for the bands of twine. Sales literature pointed out that the grooves prevented the twines slipping off the bale when they were handled and that under average conditions the grooves saved up to twenty feet of twine for every ton baled.

Most high-density balers were owned and used by threshing contractors but alternative farmer-owned lightweight stationary balers and baling presses were made by Claas, Lorant, Nicholson and Opperman and by Daniel Ross in Scotland.

The Ross Junior baler with a 5 hp Petter petrol engine or a 5 hp three-phase electric motor had an output of about 1½ tons of hay an hour. It made rectangular 15 x 18 in hay bales weighing up to 60 lb. According to a 1943 advertisement a farmer owning a

2.5 The Jones Panther automatic twine-tying stationary balers made three bales, each weighing up to 85 lb, in one minute.

Nicholson light portable-type baling press could do his baling when he wished without having to wait for a contractor to do the work. The Nicholson press driven by a 3½ hp engine mounted on the machine made up to one ton of 60 lb hay bales in an hour.

The Opperman lightweight baler was made in the late 1940s and early 1950s by SE Opperman at Boreham Wood. Similar to the Nicholson baling press the Opperman baler, driven by multiple vee-belts from a 6 hp petrol engine or a 5 hp electric motor, would 'with reasonable team work bale up to a ton of hay in an hour'.

August Claas, who patented the Claas knotter in 1921, was making stationary straw trussers in the early

2.6 Made in Scotland the Ross Junior baler with an air-cooled Petter engine cost just under £400.

1920s. As the Claas knotter was considered superior to other designs it was also used for other makes of baler. D Lorant Ltd at Radlett in Hertfordshire imported Claas RLS lightweight stationary hay and straw presses until the outbreak of World War Two. A limited number of Claas stationary presses were made under licence at Radlett during the war years but with the conflict at an end Lorant renewed their link with Claas and imported low-density stationary presses. They were delivered to Radlett without their wheels which were fitted during the pre-delivery check at the Lorant factory.

The Claas RLS press, which made 20-50 lb trusses of hay and straw, was used either behind a threshing drum or in the field with a self-feed elevator. An optional rear chute extension could be used to push the trusses up to 20 ft in height and 60 ft away from the press to load a trailer or stack the trusses in a barn.

Pick-up Balers

Wire was used to bind the bales on most early high-density American pick-up balers. A seat was provided on one side of the bale chamber for a man to thread two wires round each bale and twist the ends together. The first New Holland Automaton Model 73 twine-tying pick-up balers, made in America in 1938, still had a seat for an operator to keep an eye on the knotters. A few New Holland Automaton balers were used on British farms in the early 1940s.

The sideways-packing Claas pick-up trusser appeared in 1934 and several hundred had been sold by the early 1940s. Twenty years later many thousands of Claas pick-up trussers and ram-type balers were in use. An agreement made between Claas and Lorant in 1948 resulted in the production of the pto-driven Model L, the engine-driven Claas Model M and the pto- or engine-driven Claas RLP lightweight pick-up presses at Radlett and sold under the Lorant name. Lorant had moved to Watford when Ransomes, Sims & Jefferies acquired the business in 1951. Production continued at Watford until 1956 when it was transferred to Ipswich and the Lorant range was re-badged as Ransomes low-density balers.

Jones Brothers, a long-established stationary baler maker at Rhosemor near Mold in Wales, made the Lion high-density pick-up baler in the mid-1940s. The engine-driven Lion baler was the first British-built automatic twine-tying pick-up baler. Principally made for large farms and agricultural contractors, the Lion was used with the tractor and baler straddling the swath. The Jones Invicta baler, introduced in 1949 and awarded an RASE Gold medal in 1950, was the world's first self-propelled, high-density, twine-tying pick-up baler. Weighing 3½ tons the Invicta, with a 42 hp four-cylinder Morris engine and twenty-four forward gears, had an output of three bales a minute.

The 1950 Jones price list included the Invicta, Panther, Tiger and Cub stationary balers and the

2.7 The New Holland 73, introduced in 1940, had a seat for a worker to keep an eye on the knotters.

Balers

2.8 The Jones Invicta was the world's first self-propelled twine-tying pick-up baler.

Lion and Minor pick-up machines. There was a choice of a 12 hp JAP tvo engine, a 15 hp Petter diesel or pto drive for the Jones Minor. It was also possible to attach a special gearbox to the baler and drive it with a flat belt from a tractor belt pulley or threshing drum. An optional 22 hp Armstrong Siddeley engine was available for the early 1950s pto-driven Jones Major pick-up baler.

Round balers were not an invention of the 1970s. The Allis-Chalmers Roto-Baler round baler, which made bales 3 ft wide and between 14 and 22 in diameter, was originally built in America and later at Essendine in Lincolnshire. The Roto-Baler, introduced to British farmers at the 1947 Royal Show, rolled the hay or straw between two bale-forming belts and wrapped them with binder twine. The ends

2.9 The ram on the Jones Minor baler ran at seventy-five strokes per minute.

2.10 The Allis-Chalmers Roto-Baler was said to bale up to 8 tons of hay in an hour.

of the twine were tucked into the bale and the absence of a knotter meant knotter problems just did not happen. However, the tractor did have to stop before wrapping the twine around each bale. The pto-driven Roto-Baler cost £545 in 1953 and the Roto-Baler with an Allis-Chalmers engine was £605.

Sales literature explained that the weather-resistant rolled bales from the Roto-Baler shed water like a thatched roof and they could be unrolled like a carpet for easy feeding or bedding. Written testimonials from American farmers included 'unrolling a bale of hay is as easy as unrolling the hall carpet' and 'My wife likes it fine too. We take a bale out to the hen house and hang it on the wall. The chickens go right after it.'

Jones Balers became a subsidiary of Allis-Chalmers in 1961. Production of the Roto-Baler and Allis-Chalmers 5000 combine remained at Essendine while haymaking equipment and pick-up balers were made at Mold. The pto-driven Jones Minor, Star and Super Star twine-tying balers were current in 1960. With a ram speed of 85 strokes a minute the Star and Super Star made six and twelve bales a minute respectively.

The Jones Star T&W and Super Star T&W superseded the Star and Super Star in the mid-1960s. The T and W denoted twine- and wire-tying models. The Star and Super Star balers were delivered to Jones dealerships with blue paintwork while Allis-Chalmers dealers received the same balers in orange livery and badged as the Allis-Chalmers 200T&W and 300T&W.

Bamfords acquired Allis-Chalmers in 1971 and although they dropped the Allis-Chalmers 5000 combine the new owners continued the production of Jones and Allis-Chalmers pick-up balers at Mold. By the mid-1970s Bamfords was making the Jones Mk10/Allis-Chalmers 505 and heavy-duty Jones Mk12/Allis-Chalmers 707 twine- and wire-tying balers.

The New Holland Automaton Model 73 twine-tying pick-up baler, with an output of 4-6 tons of 70-lb bales in an hour, was introduced to American farmers in 1938. Like other balers of the day a canvas conveyor, cross-feed auger and a nodding head packer fed the swath into the bale chamber. A telescopic connecting rod stopped the ram and nodding packer for one cycle while the bale was tied with two bands of twine. The Model 73 and improved models, including the Automaton 75 which appeared in 1943, the Model 76 which appeared in 1946 and the lightweight New Holland 77 launched in 1949, were equipped with a Producti-Meter bale counter.

The New Holland Model 66 twine-tying baler appeared in America in 1953 and the first British-built pto-driven Model 66 balers were made near Stroud in 1955. The New Holland 66 had a 13 hp petrol or a tvo JAP engine, an optional Enfield diesel engine being added in 1958. An auger carried the swath from the pick-up into a pre-compression chamber where the crop was partly folded before it passed into the bale chamber. New Holland explained that this unusual feature was an important factor in the baler's output of up to 7 tons an hour when baling hay. The infinitely variable bale length adjustment on the Model 66 baler allowed the tractor driver to make bales between 12 and 52 in long.

The new American-built twine-tying Super 77 and the wire-tying Model 77 were exhibited alongside the Model 66 at the 1956 Royal Show. Publicity material for the Model 77 explained that for added safety the twisted ends of the wire were tucked into the bale. The Hayliner 68 and Super Hayliner 68 with aluminium packer fingers superseded the Model 77 in 1959 when the optional extras for the Super Hayliner included a 16 hp Enfield diesel engine and wire twisters.

The Compact Hayliner 65 for small farms, the Super Hayliner 68 and Super Hayliner 78 were current in the mid-1960s. The Compact Hayliner 65 baler with a smaller 12 x 16 in bale chamber made bales between 14 and 48 in long. Sales literature explained that the optional bale thrower driven by a small petrol engine automatically propelled bales from the bale chute into a trailer. The advantage was that 'only one man was needed to bale the crop and there was no one on the trailer to slow you down'.

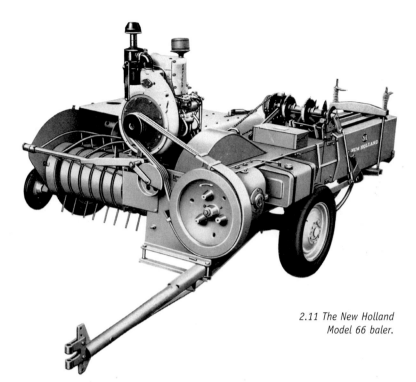

2.11 The New Holland Model 66 baler.

The streamlined, low-profile Super Hayliner 268, the Super Hayliner 278 and the 65E were the New Holland models in the late 1960s. A feature of the Super Hayliner 268 and 278 balers was their smooth auger-less sweep-tine mechanism used to feed the crop from the pick-up deck into the bale chamber. Wire twisters were optional for the Super Hayliner 268 and 278 balers. The compact twine-tying Hayliner 65E was described in sales literature as a glutton for work on hillsides and in small fields. It was also mentioned that unified threads were used throughout the machine. This thread form was used by some manufacturers for a brief period in the late 1960s before the farm machinery industry adopted metric threads.

The New Holland Super Hayliner 276 and 286 and smaller Hayliner 274, with a 70 strokes per minute ram making 14 x 18 in bales between 12 and 52 in long, were current in the early 1970s. The Super Hayliner 276 was advertised as the hungry one and the 286 as a fast, gentle giant with massive muscle to make easy work of the biggest baling jobs. Sperry New Holland pick-up balers in the mid-1970s included the 370, 376 and 386 with close-spaced pick-up cylinder tines for gathering short hay and straw and even

2.12 The New Holland Super Hayliner 268 with a 14 x 18 in bale chamber made bales between 12 and 52 in long.

higher ram speeds than the earlier Hayliner balers.

The engine-driven Massey-Harris 701 was one of the most popular pick-up balers in the late 1940s. A chain and slat elevator carried the crop from the pick-up cylinder to an auger that cross-conveyed the crop to a nodding head packer which fed the hay or straw into the bale chamber. Various engines were used for the 701 baler. Farmers at the 1953 Royal Smithfield Show had the choice of a Ferguson TED20 vaporising oil engine or a twin-cylinder Armstrong Siddeley diesel. The 701 was the only Massey-Harris baler, later designated the Massey Ferguson 701, until the launch of the Massey Ferguson 703 in 1958. The new MF703, with a 4 ft wide pick-up cylinder and self-lubricating knotters, had wooden front packer fingers that would be chopped off by the ram knife if they were mistimed and happened to be in the bale chamber at the wrong time.

The MF10, which replaced the MF701 in 1960, was similar to the smaller MF703 but with a higher output and wider pick-up cylinder. The 10 tons an

2.13 The 701 baler made three 60 lb bales in a minute.

2.14 It was claimed that the M-F15 baler could make up to 3,000 bales in a day.

2.15 The Suretie knotters on the MF124 baler launched in 1985 were claimed to tie knots 15–20 per cent stronger than those tied by conventional knotters.

hour MF15 and 11 tons an hour MF20 superseded the 703 and MF10 in 1963. Compared with earlier pick-up balers with dozens of grease nipples the MF15 and MF20 had nine greasing points, only four of which required daily attention. The MF124 was added to the Massey Ferguson pick-up baler range in 1974. The 80 strokes per minute ram ran on three hardened steel rollers and the new 'Suretie' knotters with 30 per cent fewer parts than other knotters could be used with any type of sisal or plastic twine without adjustment.

Pto- and engine-driven automatic twine-tying pick-up balers, a few with optional wire twisters for high-density bales, were in widespread use by the late 1950s. Massey Ferguson, New Holland and McCormick International balers were the most popular. Others on the market included pick-up balers made or imported by Blanch, David Brown, Jones, Salopian and Welger.

Colman & Co introduced the French-built Rival low-density straight-through pick-up baler in 1959 when AB Blanch at Crudwell imported the Rousseau low-density baler. Badged as the Blanch Model C low-density pick-up baler it was available with pto or an optional 5 hp diesel engine. Using standard binder twine the 5 tons an hour Blanch Model C made bales weighing between 12 and 44 lb.

The David Brown Albion crosswise ramming high-density pick-up baler launched in the mid-1950s was one of a kind. With the ram working at right angles to the direction of travel the baler was shorter and more manoeuvrable than most conventional pick-up balers and the surge from the fore and aft ram suffered with other balers was eliminated. Modern-looking fibreglass guards were used for the main drive units on the 8 tons an hour baler.

Salopian Engineers at Prees in Shropshire introduced the TD twine-tying pick-up baler with a rubberised canvas conveyor to carry the crop to a nodding head packer in 1951. Three versions of an improved 6 tons an hour TD Mk2 Salopian baler were launched at the 1953 Royal Show. The ET baler had a diesel engine, the PT was pto-driven and the ETP had a dual engine and pto drive arrangement. The Salopian baler with a 65 strokes per minute ram, which made hay bales weighing up to 55 lb, could also be used behind a threshing machine.

Western Machinery & Equipment Co at Ivybridge

2.16 The late 1950s Blanch Haymaster high-density straight-through pick-up baler made 3 ft wide bales weighing between 45 and 60 lbs.

2.17 The David Brown Albion baler made eight full-size hay bales in a minute.

in Devon imported Welger and Vendeuvre pick-up balers in the early 1950s. The mid-1950s Welger baler range from Germany included the AP10 and AP15 balers, the medium-density WSA trusser and a trusser attachment for combine harvesters. The pto-driven 5 tons an hour WSA was a straight-through trusser for 15 hp-plus tractors. It made 39 x 12 in bales 16 to 32 in long and weighing between 18 and 44 lb. The pto- or diesel engine-powered AP10 was a conventional pick-up model which made bales up to 40 in long in the 14 by 19 in bale chamber.

The Vendeuvre wire- or twine-tying pick-up baler was made in France. The high-capacity baler with an output of up to 10 tons an hour cost £1,050, making it one of the most expensive options on the market. The Vendeuvre, with a water-cooled diesel engine and a 56 in wide pick-up cylinder, could also be used to bale straw from a threshing machine.

By the late 1950s the Welger range included the twine-tying AP12 and AP20, the wire-tying AP30 and the WSA350 trusser. The conventional high-density AP12 pick-up baler with an air-cooled diesel engine, which replaced the AP10, was announced at the 1958 Royal Smithfield Show. The AP12 had a ram speed of

2.18 The first Salopian balers were made in 1951.

2.19 The pto- or engine-driven Welger WSA350 baled up to 5 tons of hay in an hour.

85 strokes per minute, somewhat higher than that of similar machines, and made bales 20 to 40 in long weighing 40-65 lb.

The pto-driven WSA350 medium-density baler, suitable for a 15-20 hp tractor, had a 39 x 12 in bale chamber and made bales between 16 and 32 in long and 18-44 lb in weight. Sales literature suggested that the high-density AP20 and AP30 balers with air-cooled diesel engines and outputs of up to 12 tons an hour in hay were ideal machines for farm contractors.

The Welger AP50 high-density baler launched in 1963 with an even higher ram speed of 90 strokes per minute and twine knotters or wire twisters had a claimed output of up to twenty bales per minute.

2.20 When hitched to a small tractor the Vendeuvre pick-up balers could be used at speeds of 2½–4 mph.

2.21 A bale thrower was an optional extra for the Welger AP63 high-density baler.

Optional equipment included a bale thrower with its own 7½ hp engine used to load a high-sided trailer towed behind the baler. A sales leaflet explained that by adding a bale thrower the tractor driver would be master of the whole operation of baling and loading without further assistance. The only snag was that the thrower put 30 per cent fewer bales into the trailer compared with loading the same trailer by hand.

High-density Welger balers with the option of twine knotters or wire twisters current in the mid-1970s included the AP45, the AP61 and the AP71. The AP71 with twine knotters produced bales up to 70 lb in weight and bales made by the wire-tying AP71 weighed up to 90 lb. The AP45, AP61 and AP71 were imported throughout most of the 1970s by Welger GB which by 1977 had added the string- and wire-tying AP52 to its range.

Claas was making pto-driven high- and low-density balers and trussers in the late 1950s. Trusses made by the low-density LD100 at a rate of up to 4 tons an hour were 14 x 39 in cross section and 12-24 in long. The bales from the 4-7 ton per hour Claas HD high-density baler with a standard 14 x 18 in bale chamber produced bales 16-40 inches in length. The HD baler cost just under £700 and the low-density model was £450.

New models from Claas announced in 1963 included the Maximum high-density trusser made at Harsewinkel to keep up with competition from other manufacturers. The Maximum, with an output of 10 tons an hour in straw, had a pair of horizontally opposed augers to feed the crop from the pick-up to a swinging ram, which folded the hay endways into the bales. The Maximum was still in production when the Magnum and Markant were launched for the 1967 harvest. The Magnum was a swing ram pick-up trusser and the Markant was the first conventional ram-type Claas pick-up baler.

The Markant, which cost £750 and had grooves to retain the twine in position, made bales 16-43 in long weighing 20-70 lb. With an optional rear bale chute and trailer hitch the bales could be direct loaded on to a trailer and stacked up to six bales high by hand.

The new high-density Constant and Trabant for 30 and 20 hp tractors respectively and the 20 tons an hour Dominant launched in 1969 were included in the eight models of baler in the 1970 Claas price list. Other 1970s balers made at Harsewinkel included the Markant, Maximum, Magnum, Medium and the LD80/100. The low-density swing ram LD80/100 balers for tractors of at least 15 hp had an output of 8 tons an hour.

Claas, which discontinued its trussers in 1973,

2.22 The swing-ram Claas Maximum baled up to 12 tons of hay in an hour.

2.23 The optional chute extension for the Claas Markant loaded the bales on to a trailer.

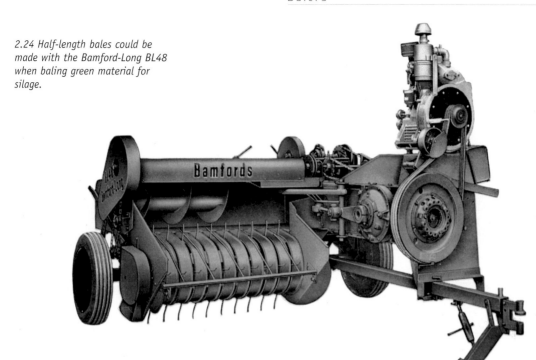

2.24 Half-length bales could be made with the Bamford-Long BL48 when baling green material for silage.

made Constant, Dominant and Markant pick-up balers at Harsewinkel throughout the 1970s and the first Claas Rollant 85 round baler appeared in 1977.

Outputs in the region of 10 tons an hour with a pick-up baler were typical in 1954 when Bamfords entered the pick-up baler market with the Bamford-Long BL60. Built under licence from the Long Manufacturing Co in America the BL60 was the first of many Bamford balers made at Uttoxeter over the next thirty years. The 10 tons an hour BL60 pick-up baler with an Enfield diesel engine cost £800. However, it was too expensive for the smaller farm and Bamfords introduced the cheaper 7 tons an hour BL48 in 1955. Prices for the pto-driven BL48 with twine knotters started at £670 with engine drive and wire twisters as optional extras.

The BL61, with a diesel engine or pto drive and optional wire twisters, replaced the BL60 at the 1959 Royal Smithfield Show and in 1960 Bamfords added the BL30 with a 12 x 16 inch bale chamber for the smaller farm. Advertised as the baler that could make every farmer independent, the pto-driven twine-tying BL30 with the ram running on ball bearings made 12-36 in long bales at a rate of five bales a minute.

Optional equipment for the pto-driven BL50, which replaced the BL48 in time for the 1964 haymaking season, included wire twisters and a 48 or 58 in wide pick-up. New models of Bamford pick-up baler appeared at regular intervals, the BL58 with the ram speed of 80 strokes a minute superseding the BL50 in 1967 and the 'luxury version' Super 58 with closer spaced pick-up tines appearing in 1969. The BL35, which replaced the BL30 in the same year, was advertised as the new economy baler with a big performance of five bales a minute.

All pick-up balers have various safety devices including shear bolts, overload slip clutches and vee-belt drives to protect the pick-up cylinder, packers, ram, knotters and needles from accidental mechanical damage. Some models, including the Bamford BL50 and BL58, were mainly protected with overload slip clutches so that little time would have to be spent replacing broken safety shear bolts.

Four Bamford balers, including the BL58 and Super BL58 with yellow paintwork, were current in 1971 when Bamfords acquired Allis-Chalmers but within a year had been replaced by the BL59 and Super 59 balers with red paintwork. More new models, including the Super 159, BL135 and the BL81 with a one-shot lubrication system, appeared in the mid-1970s. However, time was running out for Bamfords which went out of business in 1981.

2.25 The Bamford BL30 was fully guarded to comply with new farm safety regulations.

The American engine-driven McCormick Deering No 15 made in 1940 was the first International Harvester pick-up baler. The No 15 required two men to ride on the machine, one to feed the crop into the bale chamber and another to tie the bales with wire. The first McCormick-Deering No 50-T automatic twine-tying pick-up balers were made in America between 1944 and 1952. A flat belt transmitted power from the 14 hp four-cylinder International Harvester Cub engine to a heavy flywheel secured to the crankshaft by a brass shear key that protected the ram and needles from mechanical damage. A number of the 8,000 six tons an hour No 50-T balers including some No 50-W wire-tying models were sold to UK farmers.

The engine-driven McCormick-Deering No 55-T

2.26 The Bamford BL135 made bales 16 in wide x 12 in high and between 12 and 50 in long.

2.27 Optional equipment for McCormick International Mk2 B45 balers included a Petter engine, bale chute extension and a trailer hitch.

and 55-W superseded the No 50-T in 1953. The B-45 made at Doncaster between 1950 and 1959 and the American No 45 were the first McCormick International pto-driven automatic twine-tying balers. The B-45 was similar to the No 50-T with a cross-feed auger and packers feeding the crop into the bale chamber. Wooden guides on the four corners of the ram needed adjusting from time to time to compensate for wear and maintain the correct clearance between the ram and stationary knives. Optional extras for the B-45 included a Petter petrol or diesel engine and a silage attachment used to make half-length bales for silage.

Although straw merchants preferred heavy wire-tied bales they were not popular with livestock farmers. To meet the demand for heavy straw bales McCormick International introduced the 10 tons an hour B-55 baler in 1956 with a choice of pto- or engine-drive and twine knotters or wire twisters.

The B-46 with new adjustable spring steel ram guides superseded the B-45 in 1960. With ram speeds of up to 75 strokes per minute, 10 strokes faster than the B-45, the new baler was claimed to make nine bales 30 in long in a minute. A re-designed ram running at even higher speeds than the B-46 was a feature of the twine- or wire-tying McCormick International B-47 baler launched in 1965. A pto speed of 625 rpm was needed to maintain the 85 strokes per minute ram speed to give an output of up to 15 tons an hour in hay compared to the maximum 10 tons an hour output with the B-46 baler.

Farmers buying a McCormick International baler in 1967 had the choice of the ageing B-47 or the new 12 tons an hour economically priced No 27 pick-up baler. Sales literature explained that the new compact No 27 baler with a low centre of gravity and short baler drawbar made it ideal for small acreage farms. The 20 tons per hour International 440 baler, with a 65 in wide pick-up cylinder, and the 430, identical except for a narrower pick-up, were current in the mid-1970s.

The Case 200 sweep-feed baler, made at Racine in Wisconsin, was marketed by the JI Case Company at Slough in the mid-1960s. The Case 200 with a £698 price tag was, according to the American sales literature, cheap enough to buy if only as little as 13 acres were baled each year. With a pto- or optional 17 hp engine-drive the Case 200 baler had an output of 10 tons an hour when baling hay.

2.28 The B47 baler superseded the B46 in 1965.

2.29 Wire twisters were optional for the early 1980s low-profile International Harvester 425, 435 and 445 balers.

2.30 Additional equipment for the John Deere 214T and 214WS balers included a bale thrower, trailer drawbar and hydraulic bale tension control.

Lundell (GB) at Edenbridge in Kent imported John Deere forage harvesters in the late 1950s. From 1962, when it became part of the John Deere organisation, Lundell (GB) also sold John Deere 214 pick-up balers in the UK. The 10 tons an hour 214T and 214WS twine- and wire-tying pick-up balers had a hydraulic bale tensioning control which required no further attention after it was adjusted to suit the prevailing crop conditions.

John Deere opened a distribution centre at Langar near Nottingham in 1966 and within a year the John Deere 224T with a £675 price tag had replaced the earlier model. The new John Deere 'Multi-Luber' push-button lubrication system linked to all of the grease points in the knotter area was a standard feature on the 224T baler. John Deere pick-up balers in the early 1970s included the 224T, the 219 for the smaller farm and the 20 tons an hour 346 with the 'Multi-Luber' system and automatic hydraulic bale tensioning. By the end of the decade, the John Deere baler range included the 332 and 342 with optional wire twisters and the new high-capacity 456 and 466

2.31 Wire twisters were available for the Garnier 340 pick-up baler.

with 6 ft wide pick-up cylinders and ram speeds of 95 strokes per minute.

Less popular makes of pick-up baler in the 1970s included the Dania D-360 and the Garnier 340. The Dania, made by Dronningborg in Denmark, was marketed in Great Britain by Dronningborg UK at Norwich. Sales literature described the Dania 360 as a low-priced baler for large and small farms with a 1.5 m wide pick-up cylinder and a ram running on a combination of roller bearings and wooden guides at 80 strokes per minute.

Prices increased drastically in the mid-1970s when the 15 tons an hour Garnier 340 high-density baler was listed at £1,399. Marketed by Garnier UK the 340 baler, with a two-year guarantee and a 63 in wide pick-up cylinder, made bales up to 48 in long and up to 100 lb in weight.

2.32 Launched in 1983 the Vicon SP460 baler with the ram running at 102 strokes per minute made bales 18 in wide, 14 in high and up to 44 in long.

Into the 1980s

Although several makes of big round baler were on the market in early 1980 the conventional ram-type pick-up baler was just about holding its own against growing competition from big balers.

Conventional balers in the early 1980s included those made by Bamfords, Claas, McCormick International, John Deere, Massey Ferguson, Welger and Vicon. They were high output machines with a 5 or 6 ft wide pick-up cylinder, ram speeds of 80-100 strokes a minute and most made high-density bales. Other features included one-shot knotter lubrication, hydraulic bale tension control and optional wire twisters for farmers wanting to make the heaviest possible bales.

The 90 strokes a minute ram on the Bamford International BX9, which ran on ball bearing rollers, made bales up to 52 in long with retaining grooves for the twine. The MF2, MF3 and the MF4 balers launched in 1985 with 55, 60 and 66 in wide pick-up cylinders had a four-way bale density control system. A hydraulic ram for the pick-up cylinder was an optional extra.

Big Balers

Big round balers were already popular in Australia and America when the Howard Big Baler system and Big Bale Gripper appeared on the British farming scene in 1972. Invented by two British farmers and manufactured by Howard Rotavators, the Big Baler made rather untidy rectangular hay bales weighing up to 14 cwt and 9 cwt when baling straw. Baling had to stop for about two minutes while the three knotters tied the 5 x 8 ft long bales with heavy-duty baler twine and discharged them from the rear tailgate. The Howard Big Baler gripper attachment on a tractor front-end loader made light work of loading and stacking the bales.

The American Hesston 5600 imported by Opico and the Farmhand Vermeer big round balers with multiple flat belt bale chambers arrived in the UK in 1974. By the end of the decade the dozen or so round balers on the UK market included those made by Claas, John Deere, Hesston, Massey Ferguson, New Holland, Vicon and Welger. With the exceptions of the New Holland

2.33 Each of the five or six bales made per acre with a Howard Big Baler was equivalent to twenty-four standard-size bales.

850 with a chain and slat bale-forming chamber and the Claas Rollant bale chamber with 21 dimpled steel rollers they all had a bale chamber formed by a set of endless flat belts. Big round balers worked in much the same way as the earlier Allis-Chalmers Roto-Baler with the pick-up cylinder which straddled the swath feeding the hay or straw into the bale chamber where it was rolled into a bale. Like the Roto-Baler it was necessary to stop while spirally wrapping twine around the bale and ejecting it from the rear door of the bale chamber.

Some round balers with a multiple flat belt bale chamber had the facility to adjust the bale diameter. Others and round balers with a chain and slat or dimpled steel roller bale chamber made fixed diameter bales. The Welger RP180 with multiple flat belts made bales 5 ft wide and 6ft in diameter. The Sperry New Holland 840 and 850 with a fixed diameter chain-and-slat bale chamber made 4 ft wide x 5 ft 6 in

2.34 The Sperry New Holland 840 round baler had a fixed diameter chain and slat bale chamber.

2.35 The International No 241 Big Roll baler made hay bales weighing up to 12 cwt.

diameter and 5 ft 6 in x 5 ft 6 in bales respectively.

The International No 241 Big Roll round baler was an example of a variable size bale chamber with nine spring-tensioned flat belts making 2 ft 6 in to 6 ft diameter bales in the 5 ft wide bale chamber. The completed bales were spirally wound with heavy twine and ejected from the hydraulically operated rear tailgate.

The Rollant with fixed-size bale chamber, launched in 1977, was the first Claas round baler. The Rollant made 5 ft long and 6 ft diameter bales weighing up to 15 cwt when baling hay. It took 45 seconds for the Rollant to wrap each completed bale spirally with twine and eject it from the hydraulically opened tailgate. The Claas Rollant Rapid 56 introduced in 1988 was the world's first non-stop round baler. It picked up the crop, formed a bale, wrapped it with net or twine and ejected it

2.36 The Claas Rollant bale chamber consisted of twenty-one dimpled steel rollers.

on the move. The Rollant 56 was an ideal baler for the farm contractor but it was considered too expensive and was eventually withdrawn from the British market.

The Howard Big Baler was the only baler on the market which made square section bales that were easier and more convenient to transport and stack than round bales. The situation remained until 1978 when the first Hesston 4800 big square baler, which needed a tractor of at least 130 hp and a 1,000 rpm pto, arrived from America. The 4 ft x 4 ft 3 in cross section Hesston high-density hay bales between 4 and 8 ft long weighed up to a ton.

The Vicon HP1600 launched in 1984 was another power-hungry big square baler with the capacity to clear up to 7 acres of straw in an hour making bales 6 ft long in its 2 ft 6 in square bale chamber. Claas, Ford New Holland, John Deere, Massey Ferguson and Welger were among the companies making big square balers in the late 1980s. The Ford New Holland D1000 big baler, which needed a 90 hp-plus tractor, had a 6 ft 3 in wide pick-up cylinder, a 2 x 3 ft bale chamber and four heavy-duty twine knotters. Standard equipment included central knotter lubrication, hydraulic road brakes and road lights.

2.37 The Vicon HP1600 made between sixty and seventy high-density bales in an hour.

2.38 The twine box on the Claas Quadrant big square baler held twenty-four balls of baler twine.

The Quadrant 1200, launched in 1987, was the first Claas big square baler. The specification included a 6 ft 8 in wide pick-up cylinder, a 28 x 48 in bale chamber, 46 strokes per minute ram and the six knotters with an electric twine failure indicator in the tractor cab were suitable for sisal or plastic twine.

As the 1990s drew to a close, farmers had the choice of a wide selection of big round, big square and conventional ram-type pick-up balers. Computer technology provided the tractor driver with electronic monitoring of the bale-forming and tying cycles and push button controls to select bale length or diameter and to start the tying or net wrapping mechanism.

Bale Handling

Bales made with a pick-up baler during the 1940s and early 1950s were either left on the ground where they fell for later collection or were stacked by a man riding on a sledge towed behind the baler. Loading single lightweight trusses with a pitchfork was reasonably easy work but it was a different matter pitching wire-tied bales weighing the best part of a hundredweight on to a farm trailer.

Manned bale sledges were often made by a local carpenter and it was not the most desirable job standing on the narrow platform getting covered in chaff and dust while stacking the bales in layers of four in piles three or four high. A hand lever or foot pedal was provided to tip the platform floor and leave the bales in neat stacks on the stubble.

Few farmworkers would have complained when the boss bought an unmanned random bale sledge in the early 1960s. Chain-towed behind the baler, the bales fell into the sledge. When it was full the driver tugged on a trip rope to release jumbled heaps of bales at intervals across the field. Most sledges held about a dozen bales, and after building them into conveniently sized piles they were stacked on a trailer with a pitchfork or a bale-loading attachment on a tractor front-end loader. Most random sledges had metal floor slats to protect the bales from sharp stones, especially on flint-infested fields. It was not unusual for the odd twine to be cut while dragging the sledge across a field.

2.39 Loading bales the old-fashioned way.

2.40 Riding on a manned bale sledge was not one of the best jobs on the farm.

Some random sledges had small transport wheels for towing the sledge from field to field behind the baler, others were dismantled in a matter of minutes and stowed on top of the baler for transport.

A pitchfork was still used on some farms to load single bales on to a trailer but many farmers had one of several types of bale-loading attachment on a front-end loader. The Barford bale attachment was typical of the type with hydraulically operated side arms gripping a pile of bales while it was loaded on to a trailer. Other bale loaders went a stage further by picking up much bigger piles of bales and carrying them to the stack. The Trojan Balemaster and the Lister Take-Put had a bale carrier on a front-end loader and another on the tractor three-point linkage. The Trojan Balemaster with front- and rear-mounted bale

2.41 Random bale sledges left heaps of a dozen or more bales in heaps across the field.

2.42 The Lister Take-Put bale carrier cost £79 in the mid-1960s.

carriers was claimed to save the cost of two trailers, a tractor, an elevator and six men. The Balemaster was used either to load bales on to a trailer or build stacks up to fifteen bales high in the field.

The Lister Take-Put handling system with front- and rear-mounted bale carriers was used to cart a combined total of 24 bales, put them down where required and even stack them in a barn. Sales literature suggested that farmers could also use the Take-Put to cart hedge trimmings and the front unit, when fitted with a scraper blade, would double up as a snowplough. The early 1970s MkII Take-Put with single- or double-acting rams carried thirty-two bales at a time.

In the early 1960s some farmers chose to leave single bales on the ground and load them on to a trailer with one of the several types of pick-up bale elevator on the market at the time. It was usual for the elevator to be attached by a quick-release linkage to the side of a trailer. A pair of guide arms directed bales to a land wheel-driven elevator chain that lifted them up to a platform from where one or two men stacked the bales on the trailer. Several companies, including the Ayrshire Elevator Co, Catchpole, International Harvester, Parmiter, Twose and Vicon, made this type of elevator used to lift the bales on to a trailer to heights varying from 5 to 9 ft.

The introduction of more advanced mechanical bale sledges and loader attachments in the early 1960s made bale handling easier. The Juggler bale accumulator made by Browns of Leighton Buzzard was

2.43 Catchpole Engineering introduced their pto-driven pick-up bale and sack elevator at the 1950 Smithfield Show.

2.44 The land wheel-driven McCormick International No 43 bale loader attached to the side of a trailer lifted the bales to a maximum height of 7 ft.

2.45 The Brown Juggler accumulator took the hard work out of handling bales.

close coupled to a baler and the tractor hydraulics automatically arranged the bales in groups of flat eights and left them at intervals on the ground. The Brown Buzzard attachment for front-end loaders with hydraulically operated claws picked up each group of bales and stacked them on a trailer. The Farmhand bale accumulator, which worked in a similar way to the Brown Juggler, dropped flat eight packs of eight or ten bales on the ground for loading with a Farmhand Power Bale Fork.

The Meijer flat eight mechanical accumulator, imported by Colchester Tillage, used the weight of the bales to operate a guide rail that alternately deflected the bales into two rows of four. No hydraulics were used and the eighth bale pushed against a lever to open the rear door and release a flat eight pack of bales. Cook mechanical sledges towed behind a baler and made by William Cook Ltd at Yaxley near Peterborough included the Mini-Six and the Double Four Autosledge. The eight bale Autosledge formed two

2.46 The eight bale heaps made by the Cook Autosledge could be stacked up to ten bales high with the Cook S8 loader attachment.

2.47 Some self-emptying random bale sledges held twenty or more bales.

layers of four bales for loading with the Cook S8 squeeze type loader and the Mini-Six sledge made single piles three bales long and two high.

A dozen or so random sledges, some holding as many as twenty bales, were made in the early 1970s. However, by this time some farmers were using more sophisticated bale-handling equipment, including Claas and New Holland bale throwers, the New Holland Stackliner and the Warwick bale trailer. The bale thrower, attached to the end of the bale chamber, had a small petrol engine to drive two short high-speed endless belts. The high-speed belts gripped the bales as they left the bale chamber and threw them into a high-sided trailer towed behind the baler. Loading was completely random, as it would have been suicidal for a man to ride in the trailer.

The New Holland Stackliner Bale Wagon provided a high-speed method of collecting and stacking up to

1,200 bales in a day. The Bale Wagon picked up single bales and with two bales on the first table it was raised to place them on to a second table that held either six or eight bales. When the second table was full it was automatically raised to form a vertical tier of bales on the main load platform. The sequence continued until the Bale Wagon was fully loaded and, depending on the model, it was ready to transport 66, 88 or 104 bales to the stack. The main load platform was raised to a vertical position at the stack and driving the tractor forward left a ready-made stack of bales. Farmers handling many thousands of bales would have found the 150 hp self-propelled New Holland 1048 automatic Stackcruiser, which could load and cart 3,000 or more bales in a day, the answer to their dreams.

The Warwick bale trailer combined a side-mounted swing fork and elevator with a trailer to collect and transport a load of 96 bales. The swing fork picked up single bales and threw them rearwards on to the elevator which carried them to an operator who stacked the bales while riding on the trailer. When full the trailer was backed up to a stack, raised to the vertical and the load of bales added.

Several makes of mechanical accumulator including the Lely Baleightsom and the later Lely Cube 8, the McConnel Balepacker and the Steelfab Superbale appeared in the mid-1970s. The automatic British Lely Baleightsom accumulator, operated by the tractor hydraulic system, formed a stack of four cross-layered pairs of bales. The accumulator received the bales direct from the baler at ground level and once in the Baleightsom, every second pair of bales was automatically turned through 90 degrees. Each pair was lifted hydraulically and, with eight bales stacked in the machine, the rear doors opened to leave a stable stack of bales on the field.

The increasing popularity of big balers in the late 1970s prompted the introduction of the McConnel Balepacker used to tie and transport small bales in packs of sixteen or twenty-four. Loading and packing the Balepacker chamber was an automatic process using a sequencing valve supplied with oil from the tractor hydraulic

2.48 Brackets were supplied to attach the mid-1960s Gascoigne Rotolifter to a farm trailer.

2.49 It only took twenty minutes for the New Holland 1006 automatic bale wagon to collect a load of eighty-eight bales.

system. The bales were packed in columns of four and each group was pushed rearwards into the Balepacker chamber. A pair of modified Bamford baler knotters tied the pack with heavy-duty Polypropylene twine.

The Superbale handling system made by Steelfab at Cardiff was an alternative machine for packing small bales into packs of twenty. Unlike the McConnel system, the Superbale had a pusher mechanism that gathered randomly dropped bales and stacked four bales in a single row on the loading platform. With four bales on the loading platform it was raised through 90 degrees to stack the bales on the main platform. A heavy-duty band was placed around the centre of the stack as it moved rearwards and, with twenty bales on the banding platform, loading stopped while the band was secured around the pack of bales.

The McConnel Balepacker and the Steelfab Super Bale presented farmers with a new dilemma: how to handle these packs of bales. Using a rough terrain fork truck or three-point linkage-mounted forklift solved the problem.

2.50 The Lely Cube 8 bale accumulator automatically stacked eight bales in four cross-layered pairs.

Big square balers and big round balers prompted the arrival of new handling equipment to collect and carry big bales to the stack or barn. Big square bales were usually loaded and stacked with a forklift truck but alternative big round bale transporters, including the Krone Benac, Brockadale and Chillington, were used

2.51 Depending on the size of the bales the McConnel Balepacker tied packs of sixteen or twenty small bales with heavy-duty twine.

to pick up and carry four or five big round bales to the stacking area. Although there were very few big square bales to the acre some farmers bought a big bale sledge to reduce the time spent handling bales. The Meijer big bale accumulator, imported by Colchester Tillage, dropped groups of three big bales on the ground for loading with a fork truck. Hesston made a three-bale accumulator and a front-end loader attachment for 80 hp-plus tractors to handle big square bales weighing up to a ton.

2.52 The Krone Benac round bale wagon took five minutes to load ten bales.

2.53 Hesston made a three-bale accumulator for the 4800 big square baler.

Chapter 3
Hay and Silage Machinery

Cutter Bar Mowers

The origins of the horse-drawn reciprocating knife cutter bar mower can be traced back nearly 200 years while the first land wheel-driven tractor mowers were made in the 1920s. The reciprocating knife cutter bar was also used for a reaping machine made by Scottish parson Rev Patrick Bell in 1826 and then for Cyrus Hall McCormick's reaper, introduced to American farmers in 1834.

The Clayton & Shuttleworth American Eagle cutter bar mower won the top prize at a trial of hay mowers organised by the Royal Agricultural Society in 1857. AC Bamlett at Thirsk in Yorkshire introduced a cutter bar mower in 1866 and Bamfords of Uttoxeter, a company founded in the 1830s to make domestic stoves, introduced a cutter bar mower in 1887. The McCormick Harvesting Machine Co and Hart Massey were two of the many American companies making one- and two-horse mowers in the 1870s and by the turn of the century they were being shipped across the Atlantic to compete with British-made machines.

Albion, Bamford, Bamlett, Howards of Bedford, Ruston & Hornsby and Tullos were some of the better-known British mower manufacturers in the early 1900s. The Bamlett No 5 two-horse mower had an enclosed gearbox and a foot pedal was provided to vary the angle of the cutter bar. A 1913 advertisement pointed out that Bamlett mowers had enjoyed an unequalled fifty-year reputation for quality of material and workmanship. Wheel-driven mowers were not very efficient when cutting heavy crops or working on difficult terrain. To overcome this problem Bamlett introduced a 4 ft 6 in cut one-horse mower with a small petrol engine and pneumatic tyres in the late 1930s. This may well be why cutter bar mowers were known in some parts of the country as grass engines.

Tractor-drawn mowers had a safety breakaway mechanism in the drawbar to protect the cutter bar from overloading. Some had an adjustable spring-loaded mechanism to hold the clevis jaw in the drawbar pole, while others used a wooden peg to secure the

3.1 An early 1900s Massey-Harris horse-drawn cutter bar mower.

clevis jaw in the mower drawbar. The wooden peg was included in the mower parts list but it was cheaper to cut a replacement from the hedge. Both designs allowed the tractor and mower to part company if the cutter bar was overloaded or hit an obstruction. However, if the tractor driver wasn't fully alert he could continue on his way without the mower.

3.2 The Bamlett No 9 Double Drive mower was made with a two-horse pole or a tractor drawbar.

3.3 The Bamlett HT-1 trailed mower tool kit included two spanners, a spare finger, six spare knife sections and a special Bamlett reaper file.

Some trailed mowers had a seat for a man who lifted and lowered the cutter bar using a hand lever, while others had a rope from the land wheel-powered lift clutch to the tractor which the driver used to raise and lower the cutter bar. When turning at the headland the cutter bar was held in the semi-lift position to prevent the swath board and the outer end of the cutter bar picking up the previously cut swath.

The mid-1940s Bamlett mower range included the left- and right-hand cut No 9 double drive mower for two horses or a tractor with both wheels driving the cutter bar. A corn reaping attachment was available for some models of Bamlett tractor mower. The swath board and small cutter bar wheel were replaced with a larger outer wheel and a row of slats attached to the back of the bar across the full cutting width collected the cut crop. A seat was provided for a second man to ride on the mower and use a long-handled rake to leave the crop in bundles at intervals on the ground for hand tying with straw bands.

The early 1950s 5 and 6 ft right-hand cut Bamlett TM1 tractor-drawn self-lift and hand-lift Easilift mowers with an automatic knife head lubrication system were gear driven from both wheels. The Bamlett HT1 mechanical lift mower introduced in 1956 was a cheaper version of the TM1 with grease nipple lubrication and an optional lever to engage and disengage the drive to the cutter bar. The importance of regular oiling and greasing was stressed in the TM1 instruction booklet. Bamlett mower owners were informed that when mowing at 4 mph, a speed not to be exceeded, the crankshaft made 65,000 revolutions

3.4 Options for the Bamford 7 RTC tractor mower included a hand- or self-lift 5 or 6 ft cutter bar and steel or pneumatic-tyred wheels.

an hour and the knife 60,000 strokes to cut an acre of grass. Bamlett also made the petrol-engined Momore tractor mower in the mid-1950s.

Founded in 1871 Henry Bamford and Sons made their first cutter bar mower, the two-horse Bamford Royal No 5, in 1887. By the late 1920s the Uttoxeter-based company was making left- and right-hand cut mowers with a pole and whippletrees for two horses or a tractor drawbar. The 7R Royal horse mower with an oil bath gearbox appeared in 1945. When the 3 ft 6 in cut Bamford A2 mower was launched at the 1950 Royal Smithfield Show, Ferguson and other manufacturers were already making fully mounted cutter bar mowers. Sales literature explained that the one- or two-horse A2 mower for the smaller farm had machine-cut gears, roller bearings and a double self-aligning pitman or connecting rod.

The 7RTC 5 ft cut trailed hand-lift tractor mower made its debut in 1933 and within a couple of years Bamfords had added the 6 ft cut 7RTX with a wheel-operated self-lift mechanism to its mower range. The 7RTX cost £82 15s in 1950 with pneumatic tyres an optional extra. Both mowers were still being made in 1963 when the price of the 7RTX had risen to £127.

Harrison, McGregor & Co, founded at the Albion Works at Leigh in Lancashire in 1873, made the Albion No 5 and other horse-drawn mowers with enclosed oil bath gearboxes in the late 1890s. Following a name change to Harrison McGregor & Guest Ltd in

3.5 The Bamford A2 horse mower cost £46 in 1946.

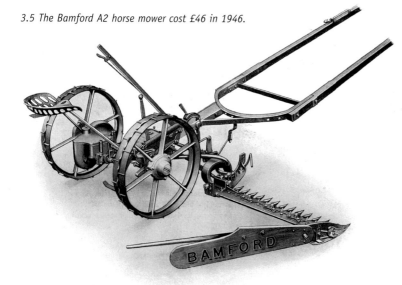

3.6 The Bamford Major semi-mounted mower had a double roller chain drive to the knife crank.

1946 the company became part of the David Brown organisation in 1955. An advertisement in the early 1940s for the two-horse Albion 16R and 16L right- and left-hand cut mowers with a seat for the horseman pointed out that the 16R and 16L were the only mowers with automatic force-feed lubrication.

The controls for the early 1940s self-lift 5, 6 and 7 ft right-hand cut Albion 8BT tractor mower were within easy reach of the tractor seat. Albion introduced the 5 and 6 ft right-hand cut Beacon self-lift tractor mower at the 1947 Royal Show. Features of the Beacon mower included a sealed oil bath gearbox, an automatic self-tightening clip to secure the connecting rod to the knife head and a spring-loaded overload latch in the drawbar. Other Albion mowers at the show included the one-horse No 12, two-horse 16R and a mower with a small petrol engine. The mid-1950s 5ft cut Albion No 20 mower had a quick-action drawbar screw jack and a spring-loaded overload safety latch on the drawbar.

Pto-driven semi-mounted mowers, including the McCormick Deering No 12B, were used in America in the late 1930s. Launched in 1937, the semi-mounted 12B was bolted on to the tractor drawbar and carried by a single castor wheel at the rear. Other semi-mounted mowers, including those made by

3.7 Harrison, McGregor & Guest made the No 18 semi-mounted Albion mower in the mid-1950s.

Albion, Bamford, David Brown, Dening, International Harvester and Massey-Harris, were popular from the late 1940s to the early 1960s.

The semi-mounted Massey-Harris 706 with two rear castor wheels had a chain drive with slip clutch protection for the 5 or 6 ft cutter bar. Like most mounted and semi-mounted mowers the 706 had a spring-loaded safety breakaway device. This allowed the cutter bar to swing backwards if it encountered a tree root or other obstruction while cutting round the edge of a field. There was no need for the driver to get off the tractor to re-engage the breakaway device as reversing the tractor automatically recoupled the cutter bar. The popular 5 or 6 ft cut semi-mounted McCormick International B 21-U with a vee-belt drive to the cutter bar had a single steel or pneumatic-tyred castor wheel. The cutter bar was protected from damage through overloading by a slip clutch in the pto shaft and a vee-belt drive to the knife drive crank.

The 5 and 6 ft cut Bamford Major semi-mounted mower with twin rubber-tyred castor wheels was suitable for most popular makes of tractor. It had a slip clutch in the pto drive and a break back mechanism for the cutter bar. A hydraulic ram to lift the cutter bar at the headland was an optional extra.

Mounted Mowers

Rear-mounted mowers date back to the late 1920s when International Harvester made the No 20 mower for direct attachment to McCormick Deering 10-20 or 15-30 tractors. The pto-driven No 20 had a hand lever to raise and lower the 5, 6 or 7 ft wide cutter bar. The frame of the 1930s No 12 McCormick Deering mower for the Farmall F-12 and similar tractors was attached to the tractor drawbar. Drive from the pto was transmitted through an oil bath gearbox to the knife crank and the cutter bar lift lever was mounted on the rear axle within easy reach of the driver. Modern in design, the 7 ft cut McCormick Deering No 12 had a work rate of 20-30 acres a day.

The 5 and 6 ft cut Ferguson mower for the TE20 tractor, introduced in 1947, was one of the first fully mounted cutter bar mowers. Originally imported from America, the Ferguson mower was made by the Pressed Steel Company in Scotland. A 3 in knife stroke was standard for cutter bar mowers but the Ferguson mower's shorter 2.8 in stroke was designed to reduce both knife speed and wear. Like other rear-mounted mowers the Ferguson had a vee-belt drive to the cutter bar and a breakaway device that allowed the bar to swing backwards when overloaded. Later Massey Ferguson mowers included the 732 with a tubular frame for the MF35 and 135; a later version of the 732 with category 1 and 2 linkage was also suitable for the MF65 and 165. An

3.8 The McCormick B23 rear-mounted mower had a spring-loaded cutter bar break-back mechanism.

3.9 Bamfords RM1 rear-mounted mowers were suitable for most three-point linkage tractors.

unusual feature of the rear-mounted MF60 was its 4¾ in knife stroke with each knife section travelling across two fingers. The sections did not travel from the centre of each finger but only far enough for the cutting edge to do its work against the ledger plate.

Bamford, Bamlett, Blanch-Lely, Busatis, Dening, Featherstone, McCormick International and Ransomes were some of the many makes of rear-mounted mowers on the market in the late 1950s and early 1960s. A cutter bar mower attachment was also made for the multi-purpose McConnel Power Arm.

The 5 ft cut Bamlett MkI Universal mower suitable for most category 1 and 2 linkage tractors had a single wheel that ran behind the left-hand tractor wheel to help the mower follow land contours. There was a choice of four knife speeds and an oil drip feed was used to lubricate the knife head. A rear land wheel was not used on the Mk2 Universal mower with two knife speeds and like earlier models the 5 ft cut Bamlett Universal Mk3, introduced in 1964, had a vee-belt drive to the knife crank.

The 5 and 6 ft cut Dening Somerset mower had a tubular frame, a wooden connecting rod and the usual safety overload mechanisms. The later Dening Somerset S46H mower, a variant of the pto-driven S46, had a hydraulic motor drive to the cutter bar. With ever-increasing tractor power, the Somerset S46E was one of several makes of rear-mounted mower with an extended drawbar and pto shaft for use with a tedder or hay conditioner.

Introduced in 1959 the 5 and 6 ft cut Bamford RM1 mounted mower was suitable for most tractors. The later 3R with a box section frame and the tubular-framed Bamford 4R also had a pto extension to drive a Wuffler and other haymaking machines. The late 1960s rear-mounted Bamford LS1 mower, like the Massey Ferguson 60, had a long knife stroke covering three fingers. Sales literature explained that this design reduced crankshaft rpm yet maintained a fast knife speed, giving a clean and fast cutting action without vibration.

High-speed mowing was a claimed advantage for the mid-1960s Blanch Lely mower with a 5 or 6 ft cutter bar. The mower also had an integral drawbar and power shaft extension for a tedder or conditioner. Blanch-Lely also made a 7 ft cut mower

3.10 The cutter bar on this Ransomes rear-mounted mower is in the semi-lift position for turning at the headland.

3.11 The double-knife Busatis mower was used at speeds of up to 7 mph.

for the North American market. The first 5 ft cut Ransomes mowers were made by Sankey in 1958 but within a year 5 and 6 ft cut Ransomes FR TM1021 mowers were built at Ipswich. Modifications saw the letters A to D added to the model number and the TM1021C had a power shaft extension and rear drawbar for mowing and conditioning at the same time.

The early 1960s Busatis double-knife mower, with two reciprocating knives held together by spring pressure to give a scissor-cutting action, was different from every other mower on the market. Suitable for any tractor with three-point linkage the Busatis mower had a work rate of up to 5 acres an hour and, depending on ground and crop conditions, the knives only needed sharpening after mowing between 10 and 24 acres.

Mid-mounted Mowers

Mid-mounted cutter bar mowers, including those made by Bamford, Featherstone and Massey Ferguson, were popular in grassland farming areas in the 1950s and 1960s. Farmers bought a mid-mounted mower because the tractor drawbar was still available for a trailer and other towed implements. The mower could also be left on the tractor throughout the entire hay- and silage-making season. Being centrally mounted on the tractor, the driver had

Hay and Silage Machinery

3.12 The early 1930s pto-driven Helvetia for Fordson tractors was one of the first mid-mounted mowers.

3.13 The Bamford mid-mounted mower was vee-belt driven from the power take-off.

an excellent view of the work and the lever for lifting and lowering the cutter bar was conveniently located near the driving seat. Drive was from a vee-belt pulley on the pto through a forward-running counter shaft to the knife drive pulley and crank pin. A break-back mechanism allowed the cutter bar to swing back if it hit an obstruction or was seriously overloaded.

Featherstone Agricultural at Boreham Wood in Hertfordshire made 5, 6 and 7 ft cut mid-mounted mowers for most popular tractors in the early 1950s. The cutter bar was raised and lowered with a hand lever or optional hydraulic ram and the vee-belt drive was designed to slip if the cutter bar was jammed with grass. Like most mid-mounted mowers the Featherstone mower had a breakaway and declutching device to protect the mower if the cutter bar encountered an obstruction. Sales literature explained that when the mower was used on a Ferguson TE20 or a similarly light tractor the breakaway device might cause the tractor to swivel round the obstacle if the cutter bar hit a solid obstruction.

The 5 ft cut Massey Ferguson 779 mid-mounted mower had a ram to raise and lower the cutter bar and a hand lever was used to adjust the pitch of the cutter bar. The 779 had the usual spring-loaded cutter bar break-back mechanism but unlike the Featherstone mower it also had a small diameter wire cable that pulled down the clutch pedal when the cutter bar hit an obstruction. It was necessary to reverse the tractor to return the cutter bar to the working position but as the wire cable held down the clutch pedal it was not quite as simple as it may have appeared to reset the cutter bar.

3.14 The MF829 mid-mounted mower was driven from the central pto on the French-built Massey Ferguson 130 tractor.

Rotary Mowers

The principle of the rotary mower is as old as the reciprocating knife cutter bar but it made little impact on the farming scene until Douglas Hayter at Spellbrook in Hertfordshire made his first tractor-drawn rotary grass cutter in 1950.

The 5 ft cut trailed rotary mower, made for a local fruit farmer, was driven from the Ferguson tractor belt pulley. A 6ft cut pto-driven rotary mower was added to the Hayter range in 1953 and this was followed by a rotary mower attached to the front axle of an E27N Fordson Major and carried on castor wheels. Drive was from the tractor belt pulley through a counter shaft and an arrangement of vee-belts to the six rotary cutting discs, each with four blades. The same Hayter rotary cutting mechanism was used for early Silorator forage harvesters.

Despite cutter bar knives requiring frequent sharpening the rival rotary mower had little impact on cutter bar mower sales until the mid-1960s when farmers had the choice of a dozen or so makes of drum, disc and flail mower.

The flail mower with free-swinging flails on a horizontal power-driven shaft was the first type of rotary mower to challenge the supremacy of the cutter bar. Within a few years drum, disc and flail rotary mowers had consigned many cutter bar mowers to the farm nettlebed.

Drum mowers had two, three or four vee-belt-driven cutting drums, each with three or four short swinging blades. Disc mowers usually had four or six gear-driven cutting discs carried on an enclosed ground level gear case. A third type of rotary mower, such as the Twose Twin-Mow, had a pair of timed gear-driven drums with two full-width fixed knives. The Twin-Mow had working speeds of up to 8 mph and as the mower was only 7 ft wide it was not necessary to swing the machine behind the tractor when taking it on the road.

Flail mowers, including those made by British-Lely, Howard, Kidd, Lister, New Holland, Ugerlose, Western Machinery and John Wilder, were popular during the mid-1960s. Flail mowers were well suited to high-speed mowing and virtually no time was lost clearing blockages. The hay crop was cut with a 5 or 6 ft wide horizontal rotor with free-swinging flails running at 800-1,000 rpm. Although the mower required a lot of tractor power, the flails served the dual purpose of cutting and conditioning the grass in one operation which made the mower popular with exponents of quick haymaking.

The Lely 5ft cut fully mounted flail mower, like

3.15 *The Hayter twin-disc Rotamower was launched in 1953.*

Hay and Silage Machinery

3.16 British Lely was one of a number of companies making flail mowers in the mid-1960s.

many of its competitors, was offset from the tractor and could be lifted and folded automatically to reduce its width to 8 ft for transport. It had category 2 linkage pins as Lely considered that a 5 ft wide flail rotor running at 1,000 rpm should be used on one of the more powerful tractors of the day. The 5ft cut mounted New Holland 30 flail mower with 32 curved reversible flails running at 1,400 rpm was made at Aylesbury.

The Howard Rotavator Co also made flail mowers for 25-55 hp tractors in the mid-1960s. The 70 in wide Howard flail mower was centrally mounted on the tractor, the 36 blade rotor ran at 1,850 rpm and a full-width rear roller was used to set the 1-4 in cutting height. The 5 and 6 ft trailed Howard Haytimer flail mower with extra wide flails had adjustable rear doors that could be set to leave the grass in an 18-60 in wide fluffed-up windrow.

High-speed rotary drum mowers with a top drive system and disc mowers with the driving gears at ground level had replaced many cutter bar and flail mowers in the late 1960s.

Drum mowers with two, three or four contra rotating drums and either two or three swinging knives, had side and rear canvas safety curtains to contain the swath and prevent the occasional flying debris doing damage to anything or anybody in the vicinity. Twin drum mowers left the cut crop in a single swath and four drum mowers left two swaths. 'Wizzle while you work' was a sales slogan for the 5 ft cut Bamford Wizzler drum mower which cost £250 when it was introduced in 1965. Multiple vee-belts drove the two pairs of twin knife contra-rotating drums at 3,000 rpm leaving the crop in two fluffed up swaths. The British Lely 5ft cut Rotomow with two large diameter cutting drums left a single swath and, like the Wizzler, it was swung back behind the tractor for transport.

Claas acquired the Bautz tractor and farm machinery business in 1969 and within a couple of years they were making the twin drum Claas-Bautz WM2 rotary mower at Saulgau in Germany. The pto-driven drums with three free-swinging blades ran at 1,900 rpm through a bevel gearbox and a separate vee-belt drive to each drum. There were four models of

Claas drum mower in the late 1970s. The twin drum WM20 and WM24 mounted mowers had 1.65 m and 1.85 m cutting widths. Three shaft-driven drums were used on the 2.1 m cut mounted WM5, and the trailed WM30, also with three shaft-driven cutting drums, cut a 2.45 m wide swath.

Farmers using disc mowers in the early 1970s had a wide choice of models including those made by Bamford, Benedict, JF, Lely, Massey Ferguson, New Holland and Tarrap. Most disc mowers were pto-driven by multiple vee-belts transmitting power to a full-width oil bath gear case with four or six gear-driven cutting discs. The two free-swinging cutting blades on each disc were timed to prevent them coming into contact with the blades on the adjacent discs.

Bamford literature explained that the cutting discs on the C460 mower were mounted on a hinged gear case. The discs ran at high speed with each disc rotating in the opposite direction to its neighbours. The 5 ft 3 in cut Bamford C460 disc mower with an output of up to 5 acres an hour required at least 30 hp at the pto. The New Holland 435 disc mower, also with four discs running at 3,200 rpm, had a working speed of up to 9 mph. The discs had three swinging blades and, like the Bamford 460, the outer disc could be fitted with a dome-shaped divider to leave a space for the tractor wheel on the next round.

The four-disc Massey Ferguson MF51 mower had the even higher cutting speed of 3,250 rpm when used at the recommended pto speed of 640 rpm. The 5 ft 5 in cut MF51, with a vee-belt primary drive and timed gear train inside the cutter bar trough, was claimed to do twice as much work as a cutter bar mower.

Three models of disc mower with four, five and six discs were made by Jones Balers at Mold. The six-disc model with a 7 ft 10 in wide cutter bar and an output

3.17 Four contra rotating cutting drums on the Bamford Wizzler left two swaths of cut grass.

3.18 Claas-Bautz WM2 drum mowers were made in the early 1970s.

of up to 6 acres an hour required a 45 hp tractor. The French Someca-Fiat four-disc Musketeer mower, described by Benedict Agricultural as a masterpiece of engineering, only needed 10 hp at the pto. The four discs, each with four swinging blades, had an unusual drive arrangement with each disc driven by a separate shaft from the mower gearbox.

3.19 Depending on conditions, the MF51 disc mower could be used at speeds of up to 10 mph.

Haymaking

Haymaking was one of the most labour-intensive jobs on the farm until the late 1800s, when the invention of the cutter bar mower led to the gradual demise of the scythe in the hay field. Turning the crop to hasten drying was done entirely by hand until the first horse-drawn overshot and undershot tedders appeared in the 1850s.

Ransomes Sims & Jefferies, Bamfords of Uttoxeter and Jeffery & Blackstone were just three of the many companies making hay rakes and tedders in the early 1900s. Jeffrey & Blackstone, which originally traded as Ashby & Jeffery and later as Blackstone & Co at Stamford, was bought by RA Lister in 1937. Ransomes, which was selling hay machinery worldwide, had made more than 50,000 Haymaker overshot tedders and hay rakes by the early 1900s. Three sizes of Ransomes Ariel and Star hay rakes were still made at Ipswich in the mid-1930s.

Several companies, including Bamfords, Dening, Lister Blackstone, Massey-Harris, and Nicholson of Newark, were selling tractor-drawn rake-bar side delivery rakes and swath turners in the 1940s and 1950s. Rake bar swath turners had a set of land wheel-driven rake bars with removable centre sections that reciprocated at almost a right angle to the swath. Side delivery raking or windrowing moved two swaths sideways to form a single larger swath on fresh ground. For swath turning, the centre rake bar sections,

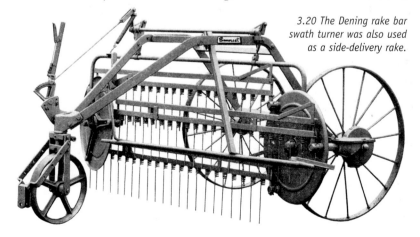

3.20 The Dening rake bar swath turner was also used as a side-delivery rake.

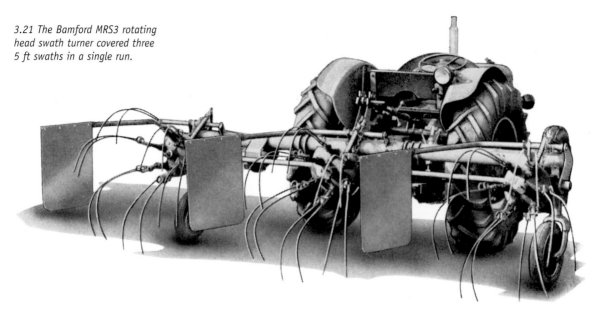

3.21 The Bamford MRS3 rotating head swath turner covered three 5 ft swaths in a single run.

secured with spring-loaded catches, were removed and the outer sections turned two separate swaths on to fresh ground. Swath turners were also used to gather previously spread grass back into two swaths.

The Bamford MRS3 and some other side-delivery rakes and swath turners had land wheel-driven rotating heads with curved spring tines, which moved or turned the swath. The rotational direction of the heads was changed with a lever. Both heads turned in the same direction for swath turning and were set to contra-rotate when putting two swaths together. Bamfords and Massey-Harris were among the several companies making rotating head swath turners and side rakes. The Massey-Harris Dickie rotating head swath turner, sold with shafts or a tractor drawbar, handled two 5 ft swaths. By the mid-1960s many of the slow wheel-driven tedders and rotating head turners had been replaced by high-speed mounted and pto-driven machines.

C Van der Lely demonstrated a horse-drawn finger wheel swath turner and side-delivery rake in Holland in 1948. Finger-wheel rakes with welded steel tines, three pneumatic-tyred wheels and a rope-operated self-lift mechanism had a high work rate but were criticised for their tendency to roll the swath into a loose rope. The finger wheels were mounted on two parallel frames and secured with quick-release catches. The six finger

3.22 Swath turning with the Bamford gearless finger wheel rake.

wheels were arranged in a continuous row for side raking, and swath turning was done with three finger wheels on each frame. Improved finger wheel machines with spring tines soon followed and within a few years many thousands had been sold.

RA Lister marketed Vicon Lely trailer rakes in the UK until 1958 when Vicon opened a depot in Ipswich and Lister lost the franchise. AB Blanch at Crudwell, which eventually became part of British Lely, also made finger wheel rakes in the

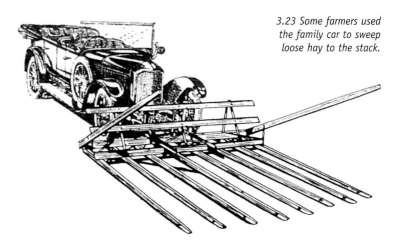

3.23 Some farmers used the family car to sweep loose hay to the stack.

3.24 The Vicon Acrobat was used for tedding, swath turning and side raking.

1950s when the range included the Blanch-Lely trailer rake, mounted Hydrake and front-mounted Forake. The mounted Vicon Acrobat with four finger wheels on walking stick-shaped frames, introduced in the early 1960s, was made for the best part of thirty years. By changing the relative positions of the frames the Acrobat could be used for tedding, swath turning and side raking.

Hay rakes were used to make a final clearance of a field after the hay cocks had been carted or loose swaths of hay had been pushed to the stack with a long wooden-tined hay sweep. Tractor-mounted rakes gradually replaced horse- and tractor-drawn trailed hay rakes in the 1950s and as more and more of the hay crop was baled the hay rake gradually disappeared from the farming scene.

Tedders hastened the drying process by lifting the hay and returning it to the ground in a loose swath. Overshot tedders with a spring-tine rotor under the hood lifted the swath over the rotor and returned it to the ground in a loosened swath for drying. The undershot tedder did not have a hood, simply lifting

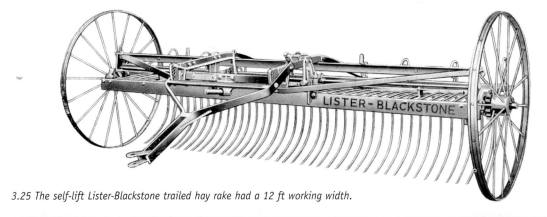

3.25 The self-lift Lister-Blackstone trailed hay rake had a 12 ft working width.

3.26 Taskers of Andover made mounted hay rakes.

3.27 The Somerset undershot tedder was made by Dening at Chard.

the swath and flicking it backwards on to the ground. Farmers replaced the shafts on some 1930s and 1940s horse-drawn tedders, swath turners and side rakes with a tractor drawbar and many were still used in the early 1950s.

Bamfords made Progress Haymaker overshot tedders at Uttoxeter in the 1870s and both horse- and tractor-drawn hay rakes and swath turners were still made there in the early 1940s. The pto-driven Bamford Wuffler W1 overshot hay tedder and conditioner, with a hood and adjustable rear doors to control the width of the tedded swath, was introduced in 1958. Farmers were advised that the best results would be achieved if the Wuffler was used directly after mowing. The wider W2 Wuffler was added in 1960 when Bamfords explained that working at speeds of 6-12 mph would leave a fluffed up swath for rapid drying and earlier baling. Wuffling had become part of the farming vocabulary when the W50 Wuffler for tedding two swaths at a time appeared in the early 1960s.

3.28 The Blanch-Lely Pheasant overshot tedder was end-towed on the road.

There were other makes of trailed pto tedder on the market in the late 1950s and early 1960s. Some were wide enough to ted two swaths at the same time. Overshot tedders included the Air-o-Tedder air-assisted hay conditioner made by Bentall-Alley Ltd, the Webb Rotary hay tedder, the Jones tedder also badged as the Allis-Chalmers Model 230, the Nicholson Robin Hood hay conditioner and the Stanhay Windrum Haymaker. The Windrum was unusual in that its high-speed square-section tedding drum created a considerable draught of air to give a fully aerated and fluffed up swath. A few companies, including David Brown, Lundell and Wilder, made a hay tedding attachment for their forage harvesters with a modified low-level chute and alternative vee-belt pulleys to reduce the speed of the flail rotor.

3.29 There were two models of the Bamford Wuffler for tedding 5 and 6 ft swaths.

3.30 There was no need to leave the swath after treatment with a Goodall Conserver as it could be carted to the silage pit without being left to wilt.

The practice of conditioning the swath by crushing or crimping freshly mown grass swaths to accelerate the drying process originated in America in the early 1950s. Trials in the UK proved that this treatment considerably reduced the drying time of the swath.

Tullos at Aberdeen launched the Goodall Grass Conserver at the 1950 Royal Smithfield Show. Unlike crimpers and roller crushers, used to crush or crimp freshly mown grass in much the same way as squeezing water from washing with a mangle, the pto-driven Goodall Grass Conserver worked in a similar way to a flail-type forage harvester. It had a rotor with swinging flails running at 1,500 rpm beneath a concave with rubber corrugations on the under surface. The rotor picked up the crop which was gently bruised against the concave and then returned to the ground in a loose swath. Sales literature recommended the same swath treatment for making silage and said that with the optional elevator on the Grass Conserver it could be used to direct load the grass on to a trailer.

Mounted and pto-driven roller crushers and crimpers were in use on UK farms by the mid-1950s.

3.31 Hay could be baled up to thirty hours earlier after using the Webb Roloflo hay crusher and crop lifter.

Some farmers used the machine as a second treatment after cutting the crop, others combined the two operations by using a mid-mounted mower and rear-mounted crusher or crimper. Roller crushers picked up the swath and squeezed the stem to release the sap. The rollers on some crushers were geared to turn faster than the forward speed of the machine. Crimpers had corrugated rollers that nipped or kinked the stems at intervals to release the sap rather than flattening them along their entire length.

Ernest A Webb at Exning near Newmarket, best known for precision seeders, fertiliser broadcasters and steerage hoes, launched the pto-driven Roloflo hay crusher and crop lifter at the 1954 Royal Smithfield Show. Advertised as a combined hay crusher, crop lifter and tedder, the three-point linkage mounted twin-roller Roloflo had a working speed of 4 mph. Tines on the bottom roller picked up the freshly mown swath before retracting to present a smooth surface that turned against the smooth upper roller to squeeze the sap from both leaf and stem. The top roller was removed when the Roloflo was used for tedding.

Several manufacturers, including Allis-Chalmers, Catchpole Engineering, McCormick International, Jones, Massey Ferguson, New Holland and the Rustproof Metal Window Co (Gehl), made hay conditioners between the late 1950s and the mid-1960s.

The French-built McCormick International F42-4 trailed and F42-5 mounted hay conditioners had vee-belt driven 4 ft 6 in wide smooth rubber rollers. The top 8 in diameter roller ran at 1,162 rpm and to improve pick-up performance the faster 5 in diameter bottom roll turned at 1,860 rpm. The early 1960s Catchpole Engineering hay conditioner with two power-driven crimping rollers was used with 5 or 6 ft cut mid-mounted mowers. Apart from the colour scheme, in the early 1960s Allis-Chalmers and Jones hay conditioners were identical, which is not surprising as by this time Jones Balers was a subsidiary of Allis-Chalmers GB Ltd. Identical sales literature explained that the machines had 50 in wide rollers, the 8 in diameter upper roll was rubber covered and the lower fluted steel roll was 6 in diameter.

The New Holland Machine Co at Aylesbury made various models of single and double swath crimpers in the late 1950s and early 1960s. The trailed 401 crimper with 73 in wide crimping rolls conditioned two swaths in a single pass. New Holland mounted and trailed 401-5 crimpers with 55 in wide rolls were single swath machines. Rear deflector doors were optional extras. Sales literature explained that the time from mowing with a crimper to baling would be less than 48 hours but without crimping it could take at least three days before the swath would be dry enough to bale.

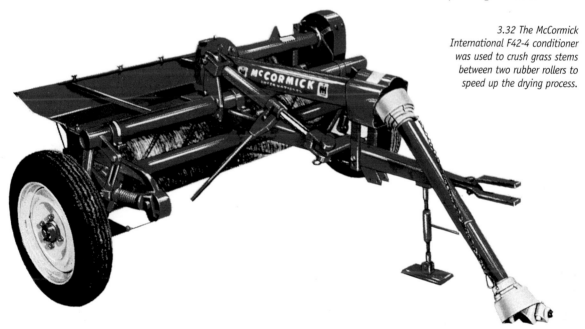

3.32 The McCormick International F42-4 conditioner was used to crush grass stems between two rubber rollers to speed up the drying process.

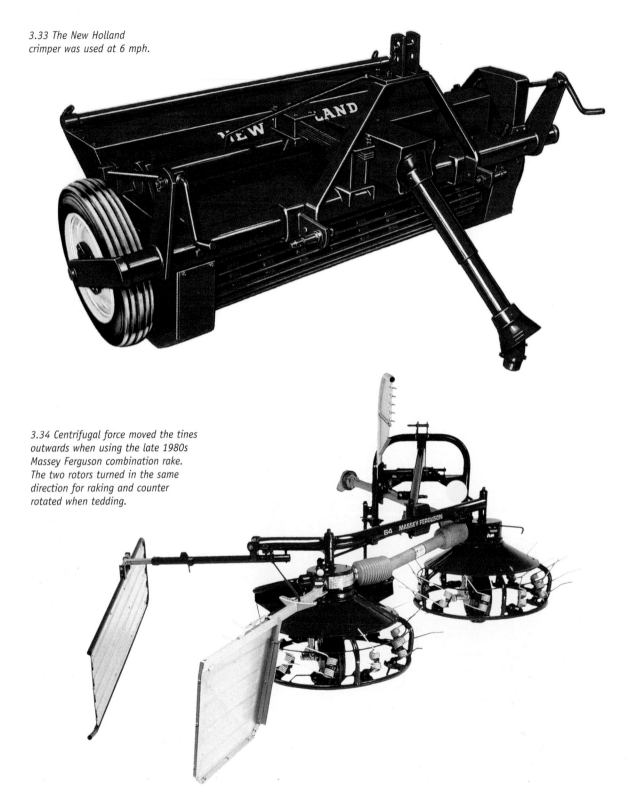

3.33 The New Holland crimper was used at 6 mph.

3.34 Centrifugal force moved the tines outwards when using the late 1980s Massey Ferguson combination rake. The two rotors turned in the same direction for raking and counter rotated when tedding.

3.35 This Vicon twin-rotor rake with 25–40 ft working widths was wide enough to match the work rate of high-speed rotary mowers used in the mid-1990s.

High-speed haymaking was a standard practice by the late 1970s. High-speed overshot tedders were popular until the introduction of a new generation of overshot and rotary tedders such as the Lely Cock Pheasant. New types of rotary tedder with six or more rotors and horizontal-wheeled tedders with spring-loaded tines to handle the swaths left by the latest models of rotary mower appeared in the mid-1980s. Even wider tedders and combination rakes were on the market in the early 1990s with working widths of 20 ft or more. Working at speeds of 10 mph or more these high-speed machines with pto-driven rotors were able to handle large swaths with minimal loss of leaf.

Hay Loaders

Grass was cut with a reciprocating knife mower at haymaking time and then moved several times with a tedder to hasten the drying process. It was left in the swath until it was dry enough to load on to wagons with pitchforks or to collect with a wooden-tined hay sweep and be taken to a stack. Hay sweeps, usually 10-12 ft wide, were either pulled along the swath by a pair of horses or pushed with the sweep attached to the front of a tractor or even the farmer's motor car (see illus. 3.23) In high rainfall areas it was usual to build hay cocks with the partly dried grass and leave them in the field until dry enough to stack.

A hay loader hitched to the back of a horse- or tractor-drawn wagon eliminated the need to load the hay with pitchforks but men still had to ride on the wagon to stack the crop. Hay loaders usually had seven or eight oscillating rake bars with curved steel tines. The rake bars were carried on four-throw crankshafts at the top and bottom of the loader trough and chain driven by the loader's large diameter steel wheels. A pair of castor wheels carried the lower end of the rake bars used to elevate the crop up to one or two men working on the trailer or wagon. The flow of hay on to the wagon was directly related to the forward speed so it was not a good idea to fall out with the horseman or tractor driver as a small increase in forward speed would soon swamp the men on the load! Later types of loader gathered the hay with a wheel-driven pick-up cylinder which passed the crop to the rake bars.

American farmers were using rake bar hay loaders in the 1920s and 1930s. An early 1920s McCormick

3.36 The pick-up cylinder and elevator trough on the Bamford CL2 hay loader were both 5 ft 8 in wide.

International catalogue included rake bar loaders with a slatted wooden trough and wooden rake bars. Massey-Harris made a similar hay loader in the early 1930s with rake bars carried on two land wheel-driven crankshafts.

Bamfords of Uttoxeter, WJ Cooper at Newport Pagnell, Dening of Chard and Salopian Engineers at Prees in Shropshire all made land wheel-driven rake bar hay loaders in the mid-1940s. The farmer-designed Fenemore Easy Six hay loader made by WJ Cooper with six rake bars and a tapered elevator trough was one of the first hay loaders on the market. Salopian Clear All hay loaders had wooden rake bars with wooden bearing blocks oscillated by three-throw crankshafts at the top and bottom of the slatted wooden elevator trough. The smaller seven rake bar model Clear All loader cost £59 5s in 1953 and the nine rake bar loader was £61 10s. The Dening Litedraft hay loader was another early 1940s oscillating rake bar loader. The lower crankshafts were chain-driven by both main wheels with over-run ratchets in both wheel hubs.

Hay loaders were gradually improved with the addition of a wheel-driven pick-up cylinder, and a steel elevator trough replaced the earlier wooden trough. As so much of the hay was baled in the early 1950s some farmers used a hay loader with a pick-up cylinder to collect green crops for silage.

Silage Machinery

Silage making in the 1940s and early 1950s was hard work. After cutting the grass with a cutter bar mower it was either loaded on to trailers with pitchforks or collected with a green crop loader and trailer and taken to the silage pit or clamp. When it was only a short distance from the field to the silage clamp many farmers used a tractor-mounted buck rake to cart the grass to the clamp.

The Wilder Cutlift, one of the first forage harvesters, appeared on a few British farms in the early 1930s. By the late 1940s some farmers with a large acreage of forage crops grown for silage used a pto-driven Fisher Humphries or an engine-driven ARM forage harvester. Others used Allis-Chalmers, Fox Rivers, International Harvester and John Deere models.

Removing silage from the pit or clamp was another

3.37 The farmer-designed Paterson buck rake was made by Taskers at Andover.

labour-intensive job. The simplest method was to let the stock help themselves but, by means of an electric fence, in a controlled way. When it had to be taken to the livestock, large blocks of silage were cut from the clamp with a suitably sharp hay knife, a chainsaw with a special cutting chain or an electric silage cutter.

Taskers of Andover made the first farmer-invented Paterson buck rakes for Fordson Major, Ferguson, Nuffield and David Brown tractors in the mid-1940s. The standard Tasker Paterson 8 ft 7 in buck rake was wide enough to collect two 5 ft mower swaths in one pass. Wider buck rakes carried up to 10 cwt of grass but plenty of front-end weight was needed to keep the tractor front wheels on the ground. About 500 yards was considered the maximum economic distance to carry grass from the field to the clamp with a buck rake. The trip mechanism built into the headstock was used to tip the grass off the tines at the clamp where the tractor wheels had the added benefit of consolidating the heap.

Buck rakes on the market in the mid-1950s included the Mil, made by Midland Industries at Wolverhampton, the Massey Ferguson New Idea and the Salopian buck rake made by the Owen Organisation. Sales literature for the 12 cwt capacity Massey Ferguson 718 buck rake for the MF35 suggested that it could also be used for moving sacks or bales and even poultry houses. The Bentall dual-purpose Culti Rake was a buck rake with an optional set of cultivator tines used to convert the Culti Rake into a spring-tine cultivator. Various attachments were available for Paterson and other makes of buck rake, including side rails for carting bales and a platform for the tines that could be used to transport churns or sacks. The Muckrake, which resembled a railway sleeper hanging from the tines, was used to clear manure in cow houses and cattle yards. Bamfords made a buck rake attachment for front-end loaders in the mid-1960s when Bamlett, Mil and other companies were advertising an optional hydraulic push-off attachment for their buck rakes.

Green Crop Loaders

Bamfords, Dening, McCormick International and Salopian were among the companies making land wheel-driven green crop loaders with steel elevator troughs in the late 1940s. Green crop loaders, used to pick up swaths of grass for silage making and vining peas, were towed behind a trailer. The early 1950s Salopian Sweeplift loader, running on steel or pneumatic-tyred wheels, had a wheel-driven pick-up cylinder and eight reciprocating rake bars in a steel elevator trough. The Bamford CL2 Universal green crop loader had a three-bar pick-up cylinder and the eight rake bars mounted on four-throw crankshafts at the bottom and top of the elevator trough elevated the crop to a maximum height of 10 ft.

3.38 The Salopian Sweeplift green crop loader was towed behind a trailer.

The McCormick Model R green crop loader, built in America between 1938 and 1953 was, with the exception of wooden rake bars, made entirely of steel. The McCormick B-R loader made at Doncaster from the late 1940s was virtually identical, the letter B denoting that it was made in Britain. The B-R had a 6 ft raking width with 20 in stroke rake bars to carry the crop to a maximum 9 ft 6 in delivery height. The man on the load was able to uncouple the spring-loaded trailer hitch with a tug on a rope. For transport, the wheel drive was disengaged with a ratchet clutch and a hinged drawbar under the loader trough was used to tow the loader backwards from field to field.

The anti-slip loader floor on the Dening Somerset green crop loader was designed to reduce the draft and to prevent choking if the rake bars were overloaded. The six rake bar Somerset loader had a 7 ft wide four-bar pick-up cylinder with the six rake bars on three-throw crankshafts lifting the crop to a height of 9 ft 6 in.

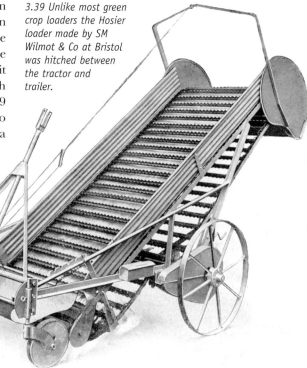

3.39 Unlike most green crop loaders the Hosier loader made by SM Wilmot & Co at Bristol was hitched between the tractor and trailer.

The late 1940s Hosier loader was one of the first green crop loaders to have an endless chain and slat conveyor running on the wooden loader trough floor. A clutch, only used when the machine was on the move, engaged the chain drive from the land wheels to the chain-and-slat conveyor. Users were advised that it was necessary to reverse the loader before disengaging the clutch and that the working speed should not exceed 3 mph. The instruction book explained that the Hosier loader could also be used for tedding by 'collecting and elevating the swath and then allowing it to float to the ground for more effective penetration by the sun and wind'.

The Bamford, Blanch-Snook, Butterley and Wilder-Steed endless conveyor loaders were towed behind a trailer. Introduced in 1952 the Bamford EL1 loader had a 5 ft wide pick-up cylinder and an endless chain-and-slat conveyor. The early 1950s Blanch-Snook green crop loader on pneumatic-tyred wheels

3.40 Launched in 1952 the Bamford EL1 crop loader with a 5 ft wide pick-up cylinder cost £180.

3.41 The makers suggested that the Petter engine on the Blanch-Snook green crop loader could be removed at the end of the season and used for other work.

was driven by a 3 hp Petter engine. The farmer-designed loader with a corrugated steel elevator floor and an endless chain-and-slat conveyor with spring tines collected the crop and loaded it into a trailer.

The later Wilder-Steed windproof green crop loader was used with its own self-emptying trailer. The wheel-driven pick-up cylinder was unconventionally situated underneath two contra-rotating endless rubber conveyor belts used to carry the crop to the trailer.

Forage Wagons

Although never popular with British farmers, forage wagons, used to pick up swaths of wilted grass and haul the grass to a silage pit, clamp or tower, were in widespread use in Holland and Germany. First imported in the early 1960s forage wagons had a front pick-up cylinder, a moving floor conveyor and a rear discharge door. Some had a side-unloading conveyor at the front of the box used to empty the load into a chopper blower for filling silage towers. An alternative type of self-unloading forage wagon or forage box without a pick-up cylinder was either hitched behind or towed alongside a forage harvester. When full the wagon was emptied into a silage pit or clamp.

Sales literature described the Blanch-Lely 3 ton LW2 loader wagon as a multi-purpose machine suitable for zero grazing, silage making, barn hay drying and even for harvesting peas and beans. The LW2 loader wagon was driven along the grass swath to collect the crop and carry it to a clamp or pit where it was unloaded from a hinged door at the back. The Vicon Hippo was basically a rear delivery manure spreader with a starting price of £543 5s. With the addition of high sides and a forage-shredding attachment, the Hippo could be used as a self-emptying forage box to cart forage-harvested grass to a silage clamp. Loader wagons and forage

3.42 The Wilder Steed loader elevated green crops between two endless rubber conveyor belts.

boxes with a front discharge conveyor were also used to fill the feed alleys and troughs in buildings used to house zero grazed cattle.

Three models of the early 1970s German-built Fahr self-loading forage wagon had load capacities ranging from 3½ - 5½ tons. Fahr forage wagons had a ten-knife cutting rotor immediately behind the pick-up cylinder for chopping the grass into short lengths.

Archie Kidd, Bamfords, Claas combine importer J Mann & Son, Bernard Krone, JF and New Holland were among the companies marketing front discharge forage boxes in the mid-1970s. The 4 ton New Holland Model 14 forage box had two beaters and a cross conveyor at the front used to fill cattle troughs or load grass into a blower used to fill silage towers. The New Holland forage box could be converted to a rear unloading forage box or a manure spreader. This was done by removing the front cross conveyor and turning the floor section through 180 degrees with the aid of a centrally mounted jacking system. With the high sides removed and the beaters re-located at the rear, the No 14 was then ready to shred and spread farmyard manure. The

3.43 The pto-driven Blanch-Lely LW2 loader wagon had a 5 ft wide pick-up cylinder.

3.44 Mowing and loading was a one-man operation with a Fahr forage wagon.

4 ton Kidd forage box could, after reversing the drive to the cross conveyor, be used to discharge grass from either side of the machine. Two front beaters above the conveyor were used for silage making and the use of a third beater was recommended to chop the grass into shorter lengths for zero grazing.

Forage Harvesters

The Wilder Cutlift, introduced by John Wilder at Wallingford in the early 1930s, was one of the first forage harvesters used to cut and collect forage crops and load them into a trailer. The pto-driven Cutlift had a 5 ft mower cutter bar, a bat reel and an endless chain and slat conveyor to elevate the cut crop into a trailer hitched at the rear. There were two models of the Cutlift in the late 1940s. The C7 Cutlift was used for crops up to 2 ft high and it was claimed that the C7-U, with an elevator reversing gear, could handle kale and other crops up to 6 ft high.

A few American forage harvesters, including those made by Allis-Chalmers, John Deere and McCormick International, were at work on UK farms in the mid-1940s. The Allis-Chalmers harvester, first made in America in 1940, had a 3 ft 4 in wide cutter bar, a reel and canvas elevator similar to that used for grain binders. The cut crop was elevated to a cylinder chopper and blown into a trailer. The McCormick International and John Deere harvesters cut the crop and a large-diameter combined flywheel chopper and paddle fan blew the chopped maize or grass into a trailer.

Harvesting green crops with a forage harvester for silage saved time and labour as the crop could be direct loaded into a trailer with high silage sides and carted to a silage pit, clamp or tower. Some farmers pulled a trailer alongside the machine while others hitched it behind the harvester. The full trailers were taken to the silage-making point where, with the trailer rear door open, the load was tipped into a pit, on to a clamp or into a dump box and blown into a tower silo.

Several production and prototype forage harvesters took part in a demonstration organised by the Royal Agricultural Society of England at Shillingford, Oxfordshire in 1950. Entries in the green crop collecting equipment section included the Wilder-Steed pick-up loader and self-emptying trailer, the Tasker Paterson buck rake and the Blanch Snook pick-up loader.

Forage harvesters at Shillingford included those made by Fisher Humphries, Opperman, Rotoscythe, Warburton-Todd and Mitchell Colman. The first pto-driven Warburton-Todd pick-up chopper-loader picked up a previously cut swath and conveyed it crosswise to a side-delivery elevator. There were two versions of the later Warburton-Todd MkII Multi-Purpose harvester with a 5 ft cutter bar. One had a side-delivery elevator for loading forage crops or peas into a trailer; the alternative model left the harvested crop in a windrow on the ground.

The prototype Opperman pick-up loader and the Rotoscythe cutter loader were both sandwiched between the tractor and a self-emptying trailer. The Opperman had a chain-driven elevator that picked up previously cut swaths and loaded them into a self-emptying trailer with a moving floor conveyor. The Rotoscythe cutter loader used two small rotary mower units with 1½ hp Villiers engines to cut the grass, and the air blast from the cutting rotors blew the harvested crop into a tipping trailer.

A reciprocating knife cutter bar and bat reel on the self-propelled Mitchell Colman tricycle-wheeled harvester cut the crop which was blown into a trailer by the engine-driven shredder and fan unit. The harvester, with a six-cylinder petrol engine, was steered by the single tricycle rear wheel. The pto-driven Fisher Humphries cutter-chopper-loader could harvest up to 8 tons of chopped material in an hour. The crop was cut with a reciprocating knife

3.45 The Wilder Cutlift was awarded RASE silver medals at the 1933 and 1937 Royal Shows.

3.46 When it was used with a side-delivery elevator the mid-1950s Blanch-Warburton multi-purpose harvester cut and loaded forage crops into a trailer.

cutter bar and reel and elevated on a short canvas conveyor to a chopper cylinder. The chopped grass was blown from the discharge chute into a trailer hitched to the harvester.

Other early 1950s forage harvesters included the ARM made by Agricultural Requisites & Mechanisations at Kidderminster, the Manea made by Johnson's Engineering at March in Cambridgeshire and the Taylor Doe silage combine. The Manea pick-up forage harvester with a Perkins P6 diesel engine or pto drive had a 4 ft 6 in pick-up and a chopper/blower loaded the harvested crop into a trailer. Pick-up chopper-loader and cutter-loader versions of the ARM harvester had a cylinder mower cutting mechanism.

Ernest Doe & Sons made a forage harvester conversion for the Massey-Harris 726 combine harvester. The cutter bar, reel and main elevator were retained on the Taylor-Doe green crop and silage combine. A chain and slat conveyor, which replaced the threshing drum, straw walkers and sieves, carried the cut crop to a trailer hitched to the back of the machine. The Wild-Thwaites straight-through cut, chop and load forage harvester, introduced in 1951, had a 30 hp Ford V8 Industrial engine. There was a choice of a 5 ft wide pick-up cylinder or cutter bar and the four-blade chopper blower, with a built-in blade-sharpening unit, blew the chopped crop into a trailer at the rear.

A similar straight-through forage harvester, made by Allis-Chalmers in America, was sold in limited numbers in the UK. The offset harvester, either driven by an Allis-Chalmers petrol engine or from the pto, had a 40 in cutter bar and bat reel or

3.47 The 20 tons an hour Manea pick-up forage harvester had a Perkins P6 diesel engine.

3.48 The David Brown Albion flail type forage harvester was introduced in 1957.

optional pick-up cylinder and a four-knife chopper blower unit.

The David Brown Albion Hurricane, Lundell, Massey Ferguson, Tarrup and Wilder Sila-Masta were some of the more popular flail-type forage harvesters in the mid- to late 1950s. The pto-driven David Brown Albion Hurricane, an in-line flail forage harvester, appeared in 1957. Like other flail-type forage harvesters the Hurricane cut the crop with high-speed swinging flails and lacerated it against a full-width shear bar. Airflow from the flail rotor blew the bruised material up a chute into a high-sided trailer. An offset version of the 40 in cut Hurricane with hydraulic cutting height control and a swivel chute for side or rear delivery into a trailer was added in 1961. Optional equipment for the Hurricane harvester included pasture-topping and haymaking kits.

Lundell at Edenbridge in Kent introduced a flail-type forage harvester with a 78 in wide flail rotor in 1956. A 60 in cut model was added and the 15 tons an hour Lundell 40 in cut harvester followed in 1958. Four vee-belts transmitted drive from the pto to a bevel gearbox and flail rotor. The gear ratio used in the bevel box depended on the make and model of tractor.

Massey Ferguson demonstrated a prototype 4 ft cut offset forage harvester in 1958. An auger carried the cut crop to a lacerating fan where it was bruised and blown into an accompanying trailer. The MF740 and the MF760 flail-type forage harvesters, launched in 1959, were made for Massey Ferguson by Lundell.

The in-line and offset models of the MF740 with 20 flails on a 40 in wide rotor which ran at 1,500-1,800 rpm had outputs of 15-20 tons an hour. A tractor with at least 35 hp at the pto was recommended for the offset MF760 with 28 flails on the 58 in wide rotor. The MF760, with swivelling top chute and an adjustable wheel track, had an output of up to 23 tons an hour. Optional equipment included conversion kits for shredding potato haulm, chopping crop residues and tedding hay.

The 10 tons an hour in-line Wilder Sila-Masta with fibreglass chute had a 4 ft 6 wide flail rotor which bruised the crop and blew it into a rear-towed trailer. Optional kits were available to convert the basic Sila-Masta to the Mulch-Masta for topping pastures and chopping up crop residues; the Straw-Masta doubled up as a straw chopper. Other kits were used to convert the Sila-Master into the Haulm-Masta and the Scrub-Masta.

The pto-driven Silorator, launched in 1952, used

3.49 The Mk III Silorator forage harvester had an output of up to 20 tons an hour when used with a small tractor.

two contra rotating Hayter-type horizontal rotary cutters to cut and bruise the crop. A vee-belt-driven fan blew the bruised grass from a swivel chute into a trailer. Sales literature for the 51 in cut MkIII Silorator launched in 1957 explained that it could also be used for haymaking, topping pastures, breaking and spreading combine straw and topping sugar beet. The work rate of the mid-1950s Silorator Dandy with a 1-7 in cutting height depended on the type of crop being harvested, the size of the towing tractor and whether the harvester was pulling a trailer. The maximum output with a big tractor was 12 tons an hour but when the Dandy was used with a small tractor the output dropped to little more than 4 tons an hour.

The Walley Gang-Mo was designed for making regular cuts of short leafy grass throughout the growing season to produce high-quality silage. The Gang-Mo loader with three 30 in wide independent wheel-driven gang mower cylinders loaded the cut grass into a trailer with a chain and

3.50 The Walley Gang-Mo green crop harvester was invented by a Cheshire farmer.

slat elevator. The Gang-Mo had a triangular-shaped tractor drawbar with an offset trailer hitch point that conveniently positioned the trailer under the elevator discharge hood.

Some farmers who cut grass for silage with a mower in the late 1950s left it in the swath to wilt and then used a forage harvester with a pick-up cylinder to load the wilted swaths into a trailer. Blanch, Fahr, Koela and Wild were among the companies making forage harvesters with a pick-up cylinder. The offset Blanch Whirlwind forager with an output of 15-25 tons an hour had a 4 ft 2 in wide pick-up cylinder and three-bladed cutter head. The in-line Wild Silage Harvester was the only semi-mounted harvester on the market at the time. A blower fan bruised the swath as it was forced through the curved fan housing and blown into a trailer. The harvester, with a pair of rear castor wheels, was mounted on the hydraulic lower lift arms used to lift the pick-up cylinder out of work at the headland.

There were at least thirty different models of flail and double-chop forage harvester on the British market in the mid-1960s. Most were in-line, offset or side-mounted flail-type harvesters, including those made by Bamford, Barford, Blanch, Kidd, Lundell, Martin-Markham, McCormick International, New Holland, Wallace and Wilder.

Double-chop forage harvesters, including those made by Archie Kidd, Lundell, New Holland and Massey Ferguson, either cut the crop with an offset flail rotor or collected previously cut and wilted swaths with a pick-up cylinder. An auger conveyed the crop to a chopper unit from where it was blown into a high-sided trailer or forage box. The chopper or cutter head, with either three or six knives and fan blades, ran at speeds of 600-1,000 rpm.

The Wallace side-mounted flail harvester was advertised as a versatile machine for grass, maize, haulm pulverising, orchard work, bracken clearance and even topping sugar beet. The side-mounted Massey Ferguson 71 harvester, hitched to a frame

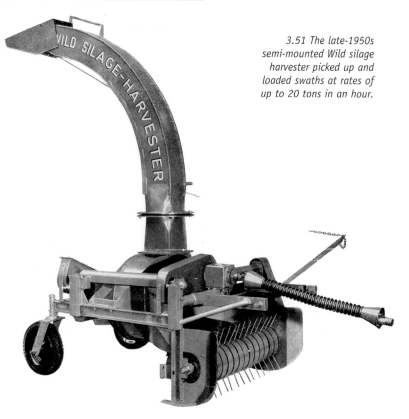

3.51 *The late-1950s semi-mounted Wild silage harvester picked up and loaded swaths at rates of up to 20 tons in an hour.*

mounted under the tractor, was pto-driven from a bevel gearbox.

There were exceptions, either in operation or design, to the basic format of flail harvesters. For the Wilder-Rainthorpe Multi-Master a stationary chopping attachment with a large hand-filled hopper above the flail rotor was also used to chop and blow straw into cattle yards. The attachment was not very user friendly as the stockman had to stand virtually under the delivery chute while loading straw into the hopper.

A top swivel chute, haymaking and maize harvesting kits and hydraulic trailer hitch were among the attachments for the 10-15 tons an hour Martin-Markham Express flail harvester. A hydraulic ram used to raise the delivery chute to give instant access for servicing the flail rotor was an unusual feature of the Express harvester.

Wilting the grass in the swath was a recommended practice for making top-quality silage and some farmers used the John Wilder Speedi-Wilt to help hasten the wilting process. Towed behind a forage harvester the early 1960s Speedi-Wilt had a large funnel-shaped hopper on wheels. Grass from the forage harvester chute was directed into the Speedi-Wilt hopper where it was fluffed up and returned to the ground in a loose, 18 in wide swath and left to wilt.

The Europa Gehl, Kidd, Lundell, New Holland, Massey Ferguson and other double chop forage harvesters gradually replaced flail-type machines in the late 1960s. The Gehl FC72, with a 72 in cut flywheel chopper harvester and cutting heights of 1½- 7 in was typical of the type. Imported by the Rustproof Metal Window Company, the Gehl double chop harvester was also used as a stationary shredder or chopper.

The 5 ft cut New Holland Model 33 Crop Chopper cut the crop with thirty-two curved flails. An auger conveyed the grass to the vertical cutter head with adjustable knives and paddles running at 975 rpm. The cutter head chopped the crop into

3.52 A swivel chute was an optional extra for the McCormick International B20-1 Gloucester forage harvester.

3.53 The side-mounted Massey Ferguson 71 flail forage harvester cost £300 in 1967.

short lengths and blew it through the swivel chute into a trailer.

Lundell introduced trailed and side-mounted double chop 42 and 60 in cut forage harvesters in the early 1960s. The Massey Ferguson 762 and 742 double chop harvesters also made by Lundell were, apart from the red paintwork, almost identical to the Lundell harvesters. The offset MF762 chopper harvester had a 5 ft wide cutting rotor with 34 curved flails and the cutter head, with either three or six knives, chopped the crop into 2-3 in lengths. Lundell, which was also importing John Deere farm equipment, was bought by the American tractor company in 1965.

Double chop forage harvesters had replaced flail machines on many farms by the early 1970s. The Gehl Chop-All, marketed by Saltney Engineering Co, was a pick-up

3.54 Some farmers used a side-mounted Ugerlose UG HM flail forage harvester in the late 1960s to top sugar beet.

3.55 The flail rotor on the New Holland Model 33 Crop Chopper ran at 1,680 rpm.

harvester with a two-speed pick-up cylinder for different types and sizes of wilted swath. John Deere forage harvesters available in the UK from the late 1960s included the 15AE double chop harvester with curved flails on the cutting rotor. The flails threw the grass into an auger trough from where it was conveyed to the cutterhead, chopped into 2 in lengths and blown into a trailer.

Precision chop harvesters, which give a more precise length of chop than double chop harvesters, appeared in the late 1960s and remain in use today. The Claas Jaguar precision chop harvester, introduced in 1969, and similar machines made by Bamfords, John Deere, Jones Balers, and New Holland were current in the mid-1970s.

Some precision chop harvesters collected previously cut swaths with a pick-up cylinder, while others cut the crop with a cutter bar and reel, a flail rotor or with a one- or two-row header when harvesting maize for silage. An auger on the pick-up deck carried the crop to the feed rolls where it was compressed into a continuous firm wad of material. The crop passed to the high-speed chopper or cutter head cylinder and shear bar where it was chopped into short lengths. The air blast created by the chopper cylinder carried the chopped material up the delivery chute into a trailer or forage box.

3.56 The early 1970s Bamford S4 precision chop forage harvester had a 25 in diameter chopper cylinder. Varying the speed and using three, six or nine blades on the chopper cylinder gave chop lengths of between ¼ in and 1½ in.

The length of chop varied from ¼ in to 2½ in or more depending on the speed of the chopper cylinder and the number of knives. Most precision chop forage harvesters had a built-in knife sharpener and some had an electric motor to rotate the swivel chute for side or rear delivery into a trailer.

The first self-propelled direct-cut forage harvesters, including the New Holland 1880 precision chop harvester, appeared on the scene in the late 1960s. Other early self-propelled models included the direct-cut Dania D5000, imported by Western Machinery & Equipment, and a precision chop harvester mounted on the self-propelled New Idea Uni-System tool frame from America.

The New Holland 1880 with Caterpillar V-8 diesel engine and the Danish-built Dania 5000 with a 165 hp Scania power unit were designed for large-scale silage-making enterprises. Both had a hydraulic transmission and hydrostatic steering, a built-in knife sharpener for the cutter head and the option of a cutter bar or pick-up cylinder.

All sorts and sizes of forage harvester from small, trailed and mounted flail harvesters to self-propelled models were on the market in the mid-1970s. Precision chop harvesters dominated the scene but in-line Hurricane harvesters made by Russells of Kirbymoorside, both in-line and offset Kidd Rotoflails and a range of Taarup flail harvesters were made for farms growing a few acres of forage crops.

Some double chop machines, including the 5 ft cut Kidd harvester for 50 hp-plus tractors and the New Holland Model 37 with a 5 ft wide flail rotor were still made. However, they were outnumbered by the wide

3.57 There was a choice of a two-row maize header, a pick-up cylinder or a cutter bar for the early 1970s New Holland 717 precision chop forage harvester.

3.58 The New Holland 1880 precision chop forage harvester was used with a cutter bar, pick-up cylinder or a maize header.

3.59 Launched in 1973, the precision chop Jaguar 60SF was the first self-propelled Claas forage harvester.

range of trailed and self-propelled precision chop harvesters on the market, including those made by Bamford, Claas, John Deere, Jones Balers, Kidd, Taarup and Vicon. Claas, a late entrant to the forage harvester market, made four variants of the Jaguar and the single-row, side-mounted 30 tons an hour Maisprinz for 35 hp tractors. The Jaguar range included 40 and 60 in cut pto-driven models, the Jaguar 60E with an 85 hp Ford or Perkins engine and the self-propelled Jaguar 60 SF with a 120 hp Perkins engine.

The New Holland 1880 and Dania D5000 were still current in the mid-1970s when self-propelled newcomers included the Hesston SP4000 with a 200 hp Caterpillar engine and hydrostatic transmission and the New Holland 1770 with a pick-up cylinder or maize head.

New Holland catered for the silage-making needs on all sizes of farm in the early 1980s when its forage harvester range included the double chop Model 37, the precision chop 717 and 770, the high capacity 880 and the self-propelled Model 1900. The double chop Model 37 with cutting flails and a pick-up cylinder or direct-cut maize attachment had feed rolls to direct the crop into the chopper cylinder and fan unit. The 717, 770 and 880 precision chop forage harvesters with a cutter bar or pick-up cylinder had an auger feed to the large diameter chopper flywheel running at speeds in the region of 1,500 rpm.

Several makes of self-propelled forage harvester, mainly used by farm contractors, and a variety of trailed double chop and precision chop harvesters were made in the 1990s. Depending on their size, self-propelled foragers with hydrostatic transmission and steering, optional four-wheel drive and an air-conditioned cab had engines in the 120-400 hp bracket. With a cutter bar, pick-up cylinder or maize header, most self-propelled precision chop harvesters, mainly used by contractors, had working widths of 10 ft or more. Remote controls to adjust the angle and direction of the delivery chute, a metal detector to prevent stray pieces of iron damaging the chopper cylinder blades and a built-in knife-sharpener for the chopper blades were all standard equipment.

3.60 The early 1990s self-propelled Claas Jaguar with a Mercedes V8 engine, hydrostatic transmission and optional four-wheel drive was used with a 10 ft wide pick-up cylinder or a six-row maize header.

Chapter 4
Sugar Beet Machinery

Sugar beet was first grown in the UK in the early 1830s when it was processed at a factory at Maldon in Essex. Another sugar factory was established at Lavenham in Suffolk in 1886 but the crop was not a success and both factories were closed. The first twentieth-century sugar factory opened at Cantley in Norfolk in 1912. Two more were built in Nottinghamshire in the early 1920s and another fifteen, including one in Scotland, opened between 1925 and 1928. Eight separate companies were involved in sugar production until 1936 when an Act of Parliament resulted in the formation of British Sugar Corporation Ltd.

About 400,000 acres of sugar beet were grown in 1948 but with fewer than 500 complete harvesters in the UK most of the crop was harvested by hand. A population of 33,000 plants to the acre was recommended. Until the early 1950s this was achieved by singling by hand the sugar beet grown from multigerm seed. The introduction of rubbed and graded seed, some with a single germ, made it possible to thin the crop mechanically. However, very few farmers were prepared to use a down-the-row thinner or gapper. A decade or so later, after plant breeders had developed single-germ pelleted seed, farmers were able to plant sugar beet with a precision spacing drill. Hand hoeing became a thing of the past and the thinners and gappers were left in the farm nettlebed.

Down-the-Row Thinners

Before the crop was thinned it was necessary to establish the plant population in order to calculate the degree of treatment required. This was done by counting the number of plants in a 100-inch length of row at random points across the field. Having established the average 100-inch plant count the instruction book was consulted to find the number and length of blades needed to reduce the plant population to the correct level.

The farmer-invented Kent horse-drawn single-row gapper made by Bentalls of Maldon was originally used to thin swedes and turnips and the tractor-mounted two-row Kent gapper was one of the first machines used to thin sugar beet plants. Root gappers were also made in the mid-1950s by Russells of Kirbymoorside and Twose of Tiverton. Two or more sets of ground wheel-driven gapping discs were supplied with the machine and the type of disc used depended on the number of plants to be removed from the row.

Most land wheel-driven thinners used a rotary or pendulum mechanism to remove some of the plants from the rows. Rotary thinners, including those made by Hudson, Ferguson and Stanhay, thinned up to six rows at a time. One or two passes were made using a number of blades of suitable size on each thinner rotor in order to reduce the plant population to the correct level. Several companies, including Catchpole, John Salmon, Root Harvesters and Vicon, made pendulum action thinners. This type of land wheel-driven thinner had cam-operated oscillating tines that removed small sections of the row as the tines moved backwards and forwards across the row.

Using a rotary or pendulum thinner was something of a gamble as it might well leave more single weeds than sugar beet plants. To overcome this problem more expensive machines guided by one or more operators appeared in the early 1960s. The farmer-designed Stanhay Silk thinning system, demonstrated at the 1961 Spring Sugar Beet Demonstration, was used on crops drilled with modified MkII Stanhay seeder units that planted three seeds abreast spaced about 9 in apart in the rows. With the plants at the true two-leaf stage the self-propelled three-row Stanhay Silk thinner with a 4 hp JAP engine and three pairs of ground-driven angled discs was driven along the rows. Three people riding on low-slung seats steered the discs with handlebars to remove the unwanted plants.

The two-row Cracknell selective hoe required

4.1 Gapping sugar beet. The men in the distance are chopping out the crop in the old-fashioned way.

4.2 Introduced in 1953, the Hudson rotary thinner was approved by the Ford Motor Company for use with its Fordson tractors.

4.3 Tines with different length blades were supplied with the Vicon pendulum thinner.

considerable concentration from the two people seated on the machine. The Cracknell had a hoe blade and a flail connected to a 12 volt electrical solenoid for each row. The operators used a cane to point at each plant to be left in the row and when the solenoid trip finger touched the cane the solenoid withdrew the hoe blade and flail just long enough to leave each selected plant growing in the row.

The even more complicated late 1960s Vicon Fähse Monomat thinner with high-speed rotary flails singled between two and six rows in one pass. An automatic touch sensor for each row located the next plant to be left growing and this activated a solenoid which withdrew the flails just long enough to leave the plant in the ground. Plant spacing could be adjusted and once set the sensor did not search for the next plant until it had

4.4 A 4 hp JAP engine was used on the self-propelled Stanhay Silk thinning machine.

4.5 There was enough room on the Martin-Markham row crop thinner platform for five people to reduce the plant population with long-handled swinging hoes pivoted on the machine.

4.6 The four people riding on this self-propelled frame singled sugar beet with hand hoes. One man (second left) steered the frame with his feet.

4.7 The Cracknell selective hoe had a work rate of up to three acres a day.

4.8 The fully automatic electro-mechanical Vicon Fähse Monomat thinned six rows at a time.

travelled the pre-set distance along the row. However, as the Monomat was unable to distinguish between weeds and plants it was important to have a weed-free crop.

Harvesters

The sugar beet crop was harvested entirely by hand in the 1920s and 1930s when the only mechanical aids were horse-drawn sugar beet-lifting ploughs, short-handled digging forks and topping hooks.

A simple horse-drawn sugar beet plough was a luxury on farms where only a few acres of sugar beet were grown but it was more usual to dig them one at a time with a short two-tined fork. Each root was knocked against the fork handle to remove

4.9 Hand topping sugar beet.

most of the soil before they were put into windrows of six or eight rows of beet to be topped with a short-handled hook. Crop debris was cleared to leave a clean space for each heap of topped roots. Clearing the trash also helped to reduce the amount of top tare carted to the factory. In cold weather it was the boy's job to cover the heaps with tops to protect them from frost. The tops had be cleared off the heaps again before the roots were hand forked on to a horse tumbrel or trailer and carted to a roadside heap. It was all hands to the fork again to load the lorry taking the beet to the factory. Sugar beet grown on farms near a railway station was sometimes carted with a horse or tractor to a siding and forked into a rail wagon.

Using trailed and tractor-mounted sugar beet lifters to loosen the roots for hand knocking and topping took some of the hard work out of the job but the soil still had to be knocked off by hand. Several types of two-row sugar beet lifter were on the market in the late 1940s and early 1950s. Some, including the Tamkin, had a single leg and share that was pulled along one side of the row to loosen the roots. Bow-type lifters, including those made by Martin, Ransomes, Stanhay and Standen, had a pair of angled shares that partly squeezed the roots from the ground.

4.10 The Tamkin root lifter eased two rows of sugar beet out of the ground.

The mid-1950s Guyco two-row sugar beet lifter, made by East Dereham Foundry in Norfolk, which looked like a sledge on cast-iron runners, loosened two adjacent rows of beet. Each pair of rows was left close together for pulling and knocking by hand. There were mounted

4.11 A two-row bow lifter was one of the attachments for the Ransomes FR toolbar.

4.12 The weight tray on the Guyco lifter had to be fully loaded on hard land.

and trailed versions of the Guyco sledge which, it was claimed, would leave the roots so clean that hand knocking would hardly be necessary.

Some farmers with a few acres of beet used a simple two-stage harvester with a separate topper and lifter cleaner made in the late 1940s and early 1950s in Scandinavia by firms such as Dameco, Mern and Roerslev.

An advertisement for the more popular Roerslev harvester explained that its topper and lifter cleaner would top and lift up to 2½ acres in a day. The horse- or tractor-drawn single-row topper sliced off the tops with a horizontal knife fixed under a feeler wheel that looked like a large and very blunt circular saw blade. Roerslev also made two-row tractor-drawn toppers. The roots were lifted with a pronged fork and passed sideways into a ground-driven cleaning cage and the cleaned roots were collected in a small hopper at the rear. A man walking behind steered the lifter cleaner with a long handle to keep the lifting fork on the row. He also emptied the hopper, leaving small heaps of sugar beet across the field. The degree of cleaning achieved by the cleaning cage depended on ground conditions, and when it was wet the roots were not always as clean as might be hoped.

The two-row Dameco topper, also with wheel-driven feeler wheels, topped up to 5 acres a day. Some farmers used an optional top-saving conveyor that left the tops in windrows to be collected for cattle feed. The single-row Dameco lifter was similar to the Roerslev but a seat was provided for the man in charge of emptying the hopper and steering the machine to keep the digging fork on the row. A later model of the Dameco lifter had a wheel-driven spinner that threw the roots into the cleaning cage.

After topping their sugar beet with one of the Scandinavian toppers some farmers used a two-row mounted Fordson lifter cleaner to harvest the crop. Introduced at the 1949 Royal Show the lifter cleaner was part of the FR implement range made at Leamington Spa. Production was transferred in 1955 to the Ransomes plough works at Ipswich. The pto-driven lifter cleaner lifted two rows of previously topped beet which were cleaned by the rotary spinner and left in a windrow. Two more rows of beet were added to this windrow on the return run ready to be hand forked on to a trailer. Suffolk tractor dealer

Cornish & Lloyd made life a little easier with the Clegg beet-loading elevator attachment for the Fordson lifter cleaner.

The Kenneth Hudson four-row tractor-mounted topper was approved by the Ford Motor Co for use with the Fordson lifter cleaner. The early 1950s Hudson topper had four independently mounted land wheel-driven feeler wheels with adjustable topping knives. A seat and steering wheel were provided for an operator to keep the feeler wheels over the rows of beet. The tops had to be removed, either by hand or with a side-delivery rake, before the crop could be lifted.

The Catchpole Engineering combined topper and top saver introduced in 1950 was based on an NIAE design. The pto-driven top saver had a vertical spiked wheel used to impale the tops as they were severed by a pair of horizontal discs. The spiked wheel carried the soil-free tops to a side-delivery elevator.

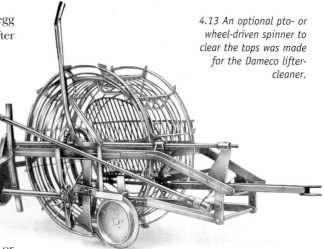

4.13 An optional pto- or wheel-driven spinner to clear the tops was made for the Dameco lifter-cleaner.

Johnson's Engineering at March, which made potato machinery, exhibited a single-row sugar beet topper at the 1950 Smithfield Show. It was mounted between the front and rear tractor wheels and the feeler wheel was chain driven from the tractor rear wheel. A pair of spring-loaded discs cut off the tops and Johnson's suggested that after the beet had been topped they

4.14 The Fordson lifter-cleaner was launched at the 1949 Smithfield Show.

could be lifted with a Johnson elevator potato digger equipped with beet-lifting shares.

A pair of bow-shaped lifters for the Lister Blackstone potato spinner appeared at the 1949 Royal Show. Priced at £9 Lister Blackstone advertised the conversion kit that replaced the potato-digging share as a novel piece of equipment that left the sugar beet in single rows with 'soil-free roots ready for hand topping'.

FW McConnel introduced the farmer-designed Silk lifter-loader, with a 3½ hp Coborn air-cooled petrol engine and a rear trailer hitch, in 1946. The crop was topped by hand in the ground and a farming magazine described the job as 'a task for four men or land girls' swinging specially designed Dutch hoes in an effortless rhythmic way to cut off the tops and throw them into a windrow in a single operation. This method of topping was claimed to leave the tops in a clean condition for feeding livestock. The topped roots, lifted with a pair of Maynard wheels, were conveyed by a rod-link cleaning elevator and then vertically to a height of 9 ft 6 in to a cleaning box from where they dropped into a trailer hitched to the harvester. An improved Birtley-Silk version of the harvester with larger lifting wheels, made at Birtley in County Durham, was demonstrated at the 1947 National Harvester Demonstration.

The two-stage Robot-Hilleshog harvester, designed in Sweden, topped four rows at a time. The tops were either collected or swept aside before lifting the crop with a two-row lifter. The lifted roots passed on to a long, inclined cleaning trough with a central auger which could be set to run at a different speeds to suit the conditions. On reaching the top of the cleaning conveyor the roots dropped into a trailer towed behind the machine. The Robot-Hilleshog harvester, made by Transplanters (Robot) at St Albans, required an operator to ride on the tricycle-wheeled pto-driven lifter to steer the lifting shares and to raise or lower them at the headland.

4.15 The Catchpole combined topper and top saver was awarded a silver medal at the 1950 Royal Show.

Two-stage harvesting soon gave way to complete harvesters, especially to those made in East Anglia by Catchpole, John Salmon and Peter Standen. In a last-ditch attempt in the mid-1950s to keep the Roerslev competitive farmers were advised that it was not difficult to convert a Roerslev single-row topper into a complete harvester. For an outlay of £8 8s a Johnson digger fork could be fitted to a Roerslev topper so that the crop could be topped and lifted at the same time.

Suffolk farmer and engineer William M Catchpole of Stanton near Bury St Edmunds made his first complete sugar beet harvester in the mid-1930s. The prototype harvester, which topped, lifted and cleaned the roots and then dumped them in heaps across the field, was awarded an RASE silver medal at the 1939 Royal Show. However, the war years intervened and several more years passed before the Catchpole was demonstrated in 1946 at a sugar beet harvesting event at Bury St Edmunds.

The pto-driven Catchpole needed a man to steer the harvester to keep it on the row. Topping was done with two horizontal discs guided by a small crawler-type feeler track running above the tops. A spinner, in front of the lifting shares, swept the tops off the row. The lifted roots were carried on a cleaning elevator to a dump hopper that was emptied at intervals across the field by the man steering the harvester. A hand-operated rake, used to clear away the tops before dumping each hopper-load of roots, was added in 1947 when William Catchpole formed the Catchpole Engineering Co.

The first John Salmon complete sugar beet harvesters, first made at Great Dunmow in Essex in 1947, were driven by a 4½ hp JAP petrol engine. The in-line harvester topped, lifted and cleaned the roots before they were side elevated into a trailer. An operator riding on the harvester used a tiller handle to keep the topper and lifting shares on the row. A choice of engine or pto drive was available for an improved John Salmon harvester launched at the 1949 Royal Show.

The first Peter Standen sugar beet harvesters made by FA Standen & Sons at Ely appeared in 1949. Frank Standen, who was making cultivators at Ely in 1906, formed FA Standen & Co in the mid-1920s to sell Austin cars and lorries. The company was appointed the local dealer for John Deere and Case tractors in 1934. Later dealerships included Massey-Harris and Ferguson. The first Standen sugar beet harvester, developed by Frank's son Peter and introduced in 1949, topped and lifted the beet before dumping them in heaps on the ground for later collection. Initially designed for the Ferguson TE20 tractor the harvester had a front-mounted feeler wheel topper with the lifting shares, cleaning conveyor and side-delivery elevator at the rear. The topper was chain driven from the rear tractor wheel and the tops were swept away before the roots were lifted. The rod-link cleaning conveyor and side elevator were driven by the pto and the shares were raised and lowered hydraulically. A second man riding on the harvester, who steered the machine with a long handle, raised

4.16 The first Catchpole complete harvester was exhibited at the 1939 Royal Show.

and lowered the topper with a hand lever. On the later dump-type harvester he also emptied the hopper at intervals across the field.

The mounted Murray harvester, introduced by the Elstree Engineering Co in 1946, had a large topping disc, lifting shares and a rotary cleaning hopper. A long-handled rake, added in 1947, was used by the man riding on the machine to clear away the tops before emptying the dump hopper. The Murray harvester, which cost £300 in 1948, was advertised for immediate delivery by the threshing machine manufacturer Tullos of Aberdeen.

The first Minns HW trailed harvesters were made in 1946 by Minns Manufacturing Co at Oxford. The pto-driven harvester topped, lifted and cleaned the beet leaving the tops and beet in adjacent windrows. The tops, removed with a pair of horizontal discs, were deflected on to a high-speed open-link conveyor that shook out any soil before returning the tops to the ground. Lifting shares put the roots on to a rod-link conveyor which carried them to a rotary cleaning cage. A cross conveyor belt, which could be adjusted to put several rows into a single windrow, returned the beet to the ground. The Minns S-SL harvester, introduced in 1948, was similar to the earlier model but a rod-link cleaning conveyor replaced the rotary cleaning cage. A side-delivery elevator loaded the crop into a trailer.

A few growers were using a complete harvester in 1949 when about 43,000 acres (10% of the UK crop) were mechanically harvested with a dozen or so makes of British and imported machines. Statistics revealed that Catchpole machines lifted more than half of the mechanically harvested crop.

Although sugar beet was not really grown in the UK until 1920, French and German farmers had grown the crop since the early 1900s. This probably explains why there were several French-built sugar beet harvesters in the country by the late 1940s. The first French Moreau harvester, sold to a Norfolk farmer in 1947, was a cumbersome two-row machine with miniature crawler tracks used to set the height of the topping discs. Shares lifted the roots which were cleaned on a long open-link conveyor. The clean beet dropped on to a cross conveyor with adjustable gates used to leave eight rows of beet in a single windrow.

Sugar beet growers in America, where more than 4,000 complete harvesters were in use by 1948, were well ahead of British farmers. International, John Deere, Marbeet and Scott Urschell were among the leading manufacturers. Although the Marbeet, tested by the NIAE, proved to be unsuitable for the heavier British farmland the other American harvesters enjoyed limited success. The International HM1, built round a Farmall H or M tractor had a picking-off table which, in difficult harvesting conditions, could be used to remove stones and clods before the roots were side elevated into a trailer.

A single-row John Deere harvester with front-mounted topper and rear-mounted lifter cleaner was working in the UK in 1946. Adjustable side conveyors left several rows of tops and beet on the ground in separate windrows. The single-row pto-driven Scott-Urschell, made by the Food Machinery & Chemical Corporation in California, was different in that it lifted the beet with the tops intact. The beet were conveyed by their tops between two endless belts to a levelling mechanism where a pair of serrated discs cut off the tops. The roots fell into a cleaning cage and were then

4.17 The International HM1 harvester took part in the 1948 autumn sugar beet demonstration.

4.18 The John Deere harvester was one of the three American machines at the 1947 National sugar beet harvesting demonstration.

side elevated into a trailer. An alternative windrowing attachment put six rows of beet into a single windrow.

The Catchpole Major Mk1 and tractor-mounted Catchpole Minor Mk1 harvesters, launched in 1951, had side-delivery elevators. An alternative rear-hitched self-emptying trailer was made for the Catchpole Major. The Major, which required a man to keep the lifting shares on the row, was one of the first harvesters to have a hydraulic ram to raise and lower the lifting shares and adjust their working depth. Top saving with the Catchpole Major harvester was achieved either by side elevating the tops into a trailer or by collecting them in a dump hopper and leaving the tops in heaps on the field.

The Catchpole Minor Mk1, suitable for Nuffield, Fordson and International tractors, had a front-mounted topper with a miniature crawler track feeler unit. The lifting shares, pto-driven cleaning cage and side elevator were mounted on the hydraulic three-point linkage. Catchpole suggested that farmers who used the tops for cattle feed might buy the Mk1 Minor lifter and cleaner after collecting the tops with a Catchpole combined topper and top saver.

British agricultural engineers were busy developing various types of trailed and three-point linkage-mounted sugar beet harvesters in the 1950s. By the end of the decade farmers could choose one of several different harvesters, including those made by T Baker & Sons, Catchpole Engineering, Fisher Humphries, GBW, the Irish Sugar Co, Salopian, Robot, John Salmon, Peter Standen and Vicon. With the exception of the earlier American Scott-Urschell, the Armer made by the Irish Sugar Co was the only mid-1950s harvester which lifted the roots before they were topped. Introduced in 1955 the tractor-mounted Armer loosened the roots with a share between the tractor front wheels. Two endless belts gripped the tops and pulled the roots from the ground and carried the complete plants up to a pair of topping discs. The roots were dropped into a cleaning cage and collected in a dump hopper. The tops fell into the top-saving hopper and were left in heaps on the field.

4.19 The Catchpole Minor Mk1 `harvester, which cost £3,760 in 1951, was suitable for a relatively modest acreage of sugar beet.

A mid-1950s Ministry of Agriculture & Fisheries report noted that 60% of the sugar beet tops grown in Britain were fed to livestock. The report stressed the importance of feeding clean tops and the Irish Sugar Co was quick to point out that this was best achieved with the Armer topping system.

The mid-1950s three-point linkage-mounted Foster-Clarke made by T Baker & Sons at Compton in Berkshire and the Fisher Humphries B-100 beet harvesters with a side-mounted topper cut the tops off the beet one row ahead of the lifting shares and cleaning cage. The Fisher Humphries B-100 also had a set of rotating rubber flickers used to sweep the tops off the row.

4.20 The Armer had separate dump hoppers for the tops and the roots.

The in-line harvester made by Salopian Engineers at Prees in Shropshire topped and lifted the same row. The ground wheel-driven feeler wheel topper with a 'specially tempered Sheffield steel knife' collected the tops and elevated them into a hopper. A man riding on the Salopian harvester dumped the tops at intervals across the field. The roots were lifted with a digging fork running alongside a large diameter cleaning cage and second dump hopper. A farming magazine noted that the Salopian harvester had 'an ingenious arrangement to intensify tractor wheel adhesion and give greater grip by putting half of the weight of the harvester on to the tractor drawbar.'

4.21 T Baker & Sons made the Foster Clarke harvester in Berkshire.

Unlike most of its competitors the Robot harvester was a lightweight affair. It had a power-driven rotary disc topper, lifting shares flanked by two angled discs, a rotary cleaning cage and a side elevator. A hydraulic ram raised and lowered the harvester into and out of work. A seat was provided for a man to steer the Robot on sloping ground but the steering gear could be locked in position when working on level fields.

FA Standen at Ely made the semi-mounted Peter Standen Junior and the trailed Universal harvesters in the early 1950s. There were side-delivery and dumper models of the Junior which topped one row ahead of the lifting shares. Suitable for Ferguson and Fordson Dexta

4.22 Transplanters (Robot) at Sandridge in Hertfordshire made the Robot harvester.

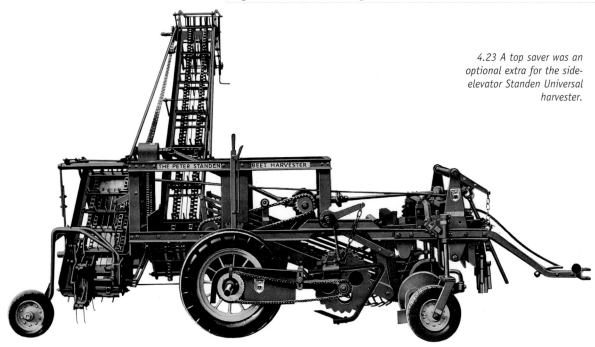

4.23 A top saver was an optional extra for the side-elevator Standen Universal harvester.

tractors, the Junior was recommended for farmers growing 5-40 acres of sugar beet on light or medium land. There was a choice of a side elevator, dumper and windrowing version of the Standen Universal harvester.

Johnson's Engineering at March, well known for its potato machinery, entered the sugar beet harvester market in the early 1950s with a potato harvester modified to harvest sugar beet.

GBW Farm Machinery, formed by Messrs Garford, Butcher and Witt, made its first sugar beet lifters in the early 1950s. The first GBW side elevator-mounted harvester with the topper unit attached to the side of the tractor was shown for the first time at the 1953 National sugar beet demonstration. The semi-mounted GBW single-row Atom 25 side-elevator harvester, suitable for most tractors with pto and hydraulic linkage, was launched at the 1957 Royal Show. An advertisement suggested that the Atom 25 would be 'unworried by rubbish'. The trailed GBW Beet King 25, introduced in 1958, had an adjustable cleaning system which could be set for working in light or heavy land and in wet or dry soil. The harvester had a built-in stone and clod remover and a self-lubricating system. The optional top saver, which collected the tops before they touched the ground, either collected the tops in a dump hopper or left them in windrows.

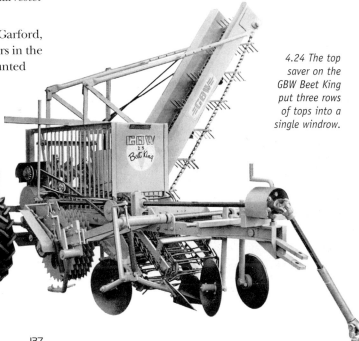

4.24 The top saver on the GBW Beet King put three rows of tops into a single windrow.

4.25 The first Catchpole Cadet harvesters were made in 1954.

GBW Farm Machinery ceased trading when the three partners went their separate ways in 1961. Boswell of Blairgowrie bought manufacturing rights for the semi-mounted Atom and the rights for the trailed GBW Atom were sold to RA Lister at Gloucester.

The Catchpole Cadet, introduced at the 1954 National sugar beet harvesting demonstration, was one of the most popular single-row harvesters with more than 800 built at Stanton in Suffolk. A seat was provided for a man to steer the Cadet but Catchpole Engineering advised prospective purchasers this would only be necessary when working on sideways sloping land. The side-elevator Cadet's land wheel-driven feeler wheel topper worked one row ahead of the lifting shares. The rod-link cleaning conveyor and side-delivery elevator were driven from the pto shaft.

Carrier, windrower and rear elevator models of the Cadet were added in the late 1950s. Carrier Cadet sales literature explained that the 18 cwt capacity hopper was large enough to hold all the roots lifted from a 20 chain length of row. There were two models of the Cadet Windrower. One had a 5 cwt hopper on the side of the harvester to collect the roots and dump them in heaps across the field. The second had an adjustable chute suspended from the side elevator that put six to eight rows of roots in a single windrow alongside the growing crop. The rear elevator Cadet, with a drawbar on the rear axle, loaded the beet directly into a trailer hitched behind the harvester.

Like the Catchpole Cadet, the single-row side elevator John Salmon 55 tanker had a side-mounted topper working one row ahead of the lifting shares. The 55 tanker cost £350 in 1960, an optional top-saving unit adding £80 to the price. A dumper version of the John Salmon 55 harvester appeared in 1964.

The Vicon Steketee, made in the early 1960s by H Vissers at Nieuw Vennup in Holland, was a side-elevator harvester with a feeler wheel topper. The far from conventional lifting unit had a pair of high-speed inclined spider wheels with four knocking weights on each wheel used to 'wriggle' the roots from the ground and throw them into a cleaning drum. Even on wet land a Ferguson 35 or Fordson Dexta was able to handle the harvester with ease when machines with conventional lifting shares were parked on the headland. However, the Steketee was not so clever in hard, dry soils as the knocking weights tended to break off the roots just below ground level. Vicon added a two-row Steketee harvester in 1964.

The mid-1950s German-built Stoll BRS2 single-row harvester marketed in the UK by Mitchell-Colman was advertised as having an output of 2½ acres in a ten-hour day working at 2 mph. The trailed, pto-driven harvester, which needed a 20 hp tractor when working on heavy land, collected the tops and roots in separate dump hoppers. Mitchell Colman entered the Stoll BRS2 harvester at the 1954 National sugar beet event when they also demonstrated an alternative method of

4.26 The Vicon Steketee was made in Holland.

topping sugar beet with a Koela chopper blower forage harvester. The tops were blown into a silage trailer for animal feed.

The Peter Standen Junior and the Beet Harvest Master were current in the late 1950s. The side elevator Junior was designed for more favourable working conditions than the more rugged Beet Harvest Master with the option of a side-delivery elevator or a holding tank. Both topped the beet one row ahead of the lifting shares but if the Beet Harvest Master had a top saver, the beet were topped two rows ahead of the shares.

Lifting wheels were in almost universal use by 1963 when Standen introduced the Rapide harvester with Maynard lifting wheels first used in the 1920s on the Maynard sugar beet lifter. The

4.27 The dump hoppers on the Stoll BRS2 harvester held 1 cwt of tops and 3 cwt of sugar beet.

4.28 The Peter Standen Beet Harvest Master was made in the late 1950s.

American Oppel lifting wheels, introduced to the UK market in 1957, were used as an alternative to shares on Catchpole harvesters. The roots were cleaned on a short rod-link conveyor and then either side elevated into a trailer or held in a 33 cwt capacity hopper. Potential buyers were advised that the Rapide, with a streamlined cover over the cleaning web and its 'squeeze and pull' lifting wheels, cut dirt tares to a minimum with no tap roots left in the ground. However, wheel lifters were not recommended for heavy, sticky land.

The John Salmon Con-Vertable, introduced in 1967, offered farmers a sugar beet harvester with either squeeze wheels or shares. Sales literature explained that the Convertible provided a quick change from lifting wheels to shares when soil conditions were wet and sticky. A squeeze wheel kit for the Con-Vertable with shares cost £45 and the

4.29 A work rate of 1½ acres an hour was possible with the Standen Rapide side-elevator harvester.

share conversion kit added £30 to the £520 basic price of the squeeze wheel harvester.

Side-delivery and tanker models of the trailed Dyson harvester were launched at the 1962 Ramsey sugar beet harvester demonstration. Made by Dyson Harvesters at Werrington near Peterborough the Dyson tanker harvester, with a 2 ton capacity hopper and a side-delivery elevator, cost £680. The side-elevator harvester cost £422 10s and an optional top saver for the tanker model added £115 to the price.

A mid-1960s economic report concluded that farmers who grew a minimum of 4 acres of sugar beet could justify the purchase of a harvester rather than relying on a contractor to harvest the crop. According to the report, depreciation and running costs would amount to about £2 16s an acre. When this figure was set against a contractor's charge of £10 an acre it was suggested that the break-even point for a farmer-owned harvester was about 4 acres of sugar beet.

Tanker sugar beet harvesters, including those made by Catchpole, John Salmon, Peter Standen and Vicon, were popular in the mid-1960s. The 1967 Catchpole Model 33 tanker with a 33 cwt capacity tank was an improved version of the earlier tanker Cadet. The tank, which emptied in ninety seconds either when stationary or on the move, had a bottom grille to return loose soil back to the ground. The Oppel lifting wheels ran on eccentric bearings and a lever was provided to adjust the distance between the wheels to suit different soil conditions and sizes of beet.

Other late 1960s Catchpole harvesters included the Super Cadet and the single-row Catchpole Sureline with the topper and Oppel lifting wheels running in line with the harvester rear wheels. Ransomes bought Catchpole Engineering in 1968 and manufacture of the orange-painted Catchpole harvesters continued at Stanton until 1971 when production moved to the Johnson's Engineering factory at March. From that date the harvesters were painted blue and sold under the Ransomes Catchpole name.

The early 1960s German Stoll harvester, together with the John Salmon Forcaster and the tanker version of the Vicon Steketee, introduced in 1961, had a forward-delivery elevator that loaded the roots into a steel hopper mounted above the tractor. No doubt the tractor driver using one of these harvesters would have appreciated the boss providing him with a pair of ear defenders. A claimed advantage of this design was that as the tank gradually filled with beet the increasing weight improved tractor wheel grip, especially in wet conditions. When full, the hopper, which held about 2 tons of beet, was tipped forward by hydraulic rams to empty the beet either into a trailer or on to a clamp. The first Forcaster harvesters had lifting shares but by the mid-1960s there was a choice of shares or lifting wheels. Lifting wheels were the only option for the late 1960s John Salmon Super Forcaster which unlike other harvesters of the day had a hydraulic depth control unit. Sales literature explained that with the dump hopper removed the frame doubled up as a tractor safety frame.

A few trailed two-row harvesters were in use by the late 1960s. They included the American two-row Farmhand tanker with Oppel lifting wheels imported by Catchpole Engineering, the Catchpole Twin Row TR2 and the Standen Beetwin. To help keep the Catchpole TR2 lifting wheels on the rows the left-hand rear tractor wheel ran in the furrow made by the Oppel wheels on the previous run. The Standen Beetwin harvester with working speeds of up to 4½ mph was launched at the 1964 Smithfield Show. The beet were topped two rows ahead of the Maynard lifting wheels and a wide rod-link cleaning conveyor carried the beet to a side-delivery elevator.

Two- and three-stage multi-row harvesters, single-row self-propelled machines and both one- and two-row complete harvesters were demonstrated at the 1968 autumn sugar beet harvester event at Caythorpe in Lincolnshire. The two-stage three-row Standen Multibeet topper, which required a 45 hp tractor, put the three rows of tops in a single windrow. The lifter-loader with three pairs of Maynard wheels lifted and cleaned the beet before side elevating them into a trailer.

Three-stage multi-row machines from France included the six-row Herriau system imported by Stanhay and the six-row Moreau marketed by Continental Farm Equipment of Norfolk. Claimed advantages for the Herriau and Moreau harvesting systems with a separate topper, lifter and cleaner loader included more efficient topping and work rates of up to 2½ acres an hour.

4.30 Hydraulic rams were used to tip the John Salmon Forecaster beet tank.

4.31 The Catchpole Twin Row TR2 harvester was made in the late 1960s.

4.32 The topper carried out the first stage of the Moreau six-row, three-stage harvesting system introduced to UK farmers in 1967.

The semi-mounted Herriau flail topper threw the leaves into a side-delivery auger, and rotary rubber beaters cleaned any remaining foliage off the roots before scalper knives completed the topping process. A semi-mounted lifter with six pairs of shares lifted the beet and passed them to a pair of power-driven horizontal rotors that dropped the roots in a narrow windrow on to the ground. For the third stage, a trailed cleaner-loader picked up the roots and passed them on to a cleaning rotor before they were side elevated into a trailer.

Sugar beet growers had a wide choice of harvesting

4.33 The Moreau lifter-cleaner left six rows of roots in a windrow.

4.34 The cleaner-loader completed the Moreau three-stage harvesting system.

systems in the mid-1970s. They included one- to four-row trailed harvesters, two- and three-stage outfits lifting three to five rows at the same time and one- to six-row self-propelled harvesters. Armer Salmon, formed when the two companies joined forces in 1974, exhibited the trailed Armer TT single-row tanker and the Armer Salmon self-propelled 21 harvester at that year's Smithfield Show. By the late 1970s the Armer range included the self-propelled 21, the trailed single-row TT tanker and the 10-acres-a-day Armer Twin Row. The TT and Twin Row with hydraulic side-tipping hoppers and the later Armer Salmon, Cougar and Cheetah harvesters followed the Armer tradition of lifting the beet before they were topped.

4.35 Garford Farm Machinery made the three- and four-row Victor harvesters in the late 1980s.

Garford Farm Machinery returned to sugar beet harvester production in 1985 when it introduced the trailed Victor harvester. The three- and four-row harvesters had a front-mounted topper/defoliator, squeeze wheel lifters and a side-delivery elevator. Later Garford harvesters included a single-row one-man machine with a front topper/defoliator, side-mounted lifting wheels and a cleaning conveyor which carried the sugar beet to a rear-delivery elevator. The harvested beet were loaded into a trailer hitched to the harvester tractor's drawbar.

Self-propelled Sugar Beet Harvesters

The mid-1940s American International HM1 (see page 133) built round a Farmall tractor was one of the first self-propelled sugar beet harvesters. The first French Moreau two-row self-propelled harvester, demonstrated to British farmers in 1952, had a 15 hp Citroen car engine at the back of the machine. The topper was similar to that on early Catchpole machines with a miniature crawler track feeler unit. The severed tops were swept away by a small side-delivery rake. Lifting shares directed the roots on to a cleaning elevator and the clean beet dropped on to a long conveyor belt running at a right angle to the direction of travel. Adjustable deflector boards spaced at intervals along the length of the conveyor belt were used to leave eight or ten rows of beet in a single windrow.

One of the first British-built self-propelled harvesters, the Dyson, made its debut at the 1963 autumn harvesting demonstration. With a 3 ton capacity holding tank or side-delivery elevator, it had a 56 hp Perkins engine, an offset topper and power-driven lifting wheels.

The single-row self-propelled Peter Standen Solobeet with a tractor skid unit was launched at a field demonstration near Ely in 1964. Standen publicity material suggested that a suitable tractor for the Solobeet might be a Massey Ferguson 35 or 135, a Ford Dexta, a Super Dexta or 3000 or an International 414 already at work on the farm. It was explained that when a farmer bought a Solobeet harvester a tractor skid unit could be collected from the farm and mounted on the new harvester. The single front wheel Solobeet with power-driven Maynard lifting wheels had a 2½ ton capacity holding tank at the rear. The Solobeet's wheels were chain driven from both sides of the skid unit rear axle. The tractor hydraulic system was used to

4.36 The Peter Standen Solobeet had a working speed of about 6 mph.

raise and lower the topper and Maynard lifting wheels.

The four-wheeled MkII Solobeet, introduced in 1968, was more stable than its three-wheeled predecessor. Other improvements included a top pulveriser working two rows ahead of the Maynard wheels. One of the lifting wheels was power driven and ran slightly faster than the forward speed to give a cleaner lifting performance.

To keep up with competition from the Solobeet, Catchpole Engineering introduced the self-propelled single-row Powerbeet with the option of a new Massey Ferguson 165 or a Ford 4000 tractor skid unit. The topper, Oppel lifting wheels and elevator carrying the beet into the rear cleaning box were in line on the Powerbeet. It took under a minute to unload the 50 cwt tank of cleaned beet into a trailer.

After Ransomes bought Catchpole Engineering in 1968 the orange-painted Ransomes Catchpole Super Cadet, 33 Tanker and Powerbeet were made at Stanton until 1971 when production was moved to March in Cambridgeshire. From that date the 33A Tanker, the improved 33B Tanker and the Powerbeet with blue paintwork were sold under the Ransomes name. The two-row self-propelled Ransomes Hunter, with some components left over from the discontinued Cavalier combine harvester, was launched at the 1974 Royal Smithfield Show. Features of the Hunter included hydraulically variable speed control in each of the three forward gears and hydrostatic steering. Electronic monitors in the cab warned the driver of problems with the beet conveyors or elevators.

John Salmon introduced the single-row self-propelled Model 21 tanker harvester in the late 1960s. Publicity material explained that most medium-powered tractor skid units were suitable for the Model 21. It was only necessary to remove the lift arms, mudguards and rear wheels before attaching the tractor to the harvester. John Salmon explained that at the end of the season it took less than three hours to make the tractor ready for other work on the farm. The topper and powered lifting wheels were located outside the harvester rear wheel and the lifted beet were cleaned as they were elevated into the holding tank centrally mounted over the rear wheels. Optional equipment for the Model 21 included a top saver, a

4.37 At the end of the harvest season the tractor could be removed from the Catchpole Powerbeet and used for other farm work.

4.38 The Ransomes Hunter with a 4½ ton holding tank could harvest up to 1½ acres an hour.

pre-topper/cutter for bolters and a weather cab.

Two-row self-propelled harvesters were gaining popularity in the 1970s. However, the big breakthrough in high-speed multi-row harvesting arrived in the UK in 1974 when a Norfolk farmer bought a French-built six-row self-propelled Moreau harvester. The first six-row self-propelled Matrot side-elevator harvester, imported from France a year or two later was similar to the Moreau with rear-wheel steering and an articulated chassis.

Various types of harvester from continental Europe dominated the 1983 autumn sugar beet demonstration. Visitors were able to compare self-propelled six-row machines made by Herriau, Kleine, Matrot, Moreau, Riecam and Vervaet. Two-stage six-row Herriau, Kleine and Standen harvesters were also at the event and Ransomes demonstrated a two-row trailed Tim harvester from Denmark. A side elevator Standen Rapide was the only single-row harvester on the field.

Six-row self-propelled sugar beet harvesters with hydrostatic steering, four-wheel drive and four-wheel steering gradually replaced complete and multi-stage trailed machines in the late 1980s. The big six-row machines with 250 hp-plus diesel engines had high capacity tanks and topping, lifting and cleaning mechanisms driven by hydraulic motors. Seated in a heated cab with tinted glass, electronic performance monitoring and joystick controls, the driver used the automatic row-finding system to keep the harvester on the row.

Farmer-owned sugar beet harvesters gradually disappeared in the 1990s and since then agricultural contractors now harvest virtually all the British sugar beet crop with high output six-row machines.

Cleaner Loaders

Loading sugar beet on to the lorries taking them to the local sugar beet factory was originally done by hand with sugar beet forks. Hand loading invariably resulted in too much soil being left on the roots as they were forked into the lorry. Carting soil to the factory cost

4.39 This six-row Cebeco harvester has four-wheel hydrostatic drive, four-wheel steering, a 343 hp DAF engine and the tank holds 16 tons of sugar beet.

money and the sugar company did not pay for the soil. Provided that the farmer had a front-end tractor loader with a root bucket, the introduction of the cleaner loader in the early 1950s made hand forks obsolete.

Most British sugar beet harvester makers were selling cleaner loaders in the early 1950s. There were two types, one with rotating cleaning cage and elevator, the other with a rod-link cleaning conveyor. Most loaders were driven by a small petrol engine although there were exceptions such as the mounted and pto-driven cleaner loaders made by Catchpole and John Salmon.

The Todd rod-link cleaner loader, made by Root Harvesters at Peterborough, took its name from one of the company directors. Todd cleaner loaders ranged from the 3 hp farm model with a loading height of 11 ft and an output of about 30 tons an hour to the contractor model with a 7 hp engine and loading rates of up to 30 cwt a minute.

Early 1960s Catchpole rod-link cleaner loaders included a free-standing model with a BSA petrol or Petter diesel engine and a pto-driven loader mounted on the three-point linkage. The 30 tons an hour Catchpole Lotare with a barrel cleaning cage had a 3.35 hp Lister diesel engine. John Salmon's Universal cleaner loader with a small petrol engine loaded up to 40 tons of beet in an hour. The Lister cleaner loader with an air-cooler Petter diesel or BSA petrol engine had a hand-operated winch which set the loading height anywhere between 8 and 14 ft.

Cleaner loaders with bigger capacity hoppers and increased loading heights appeared in the 1970s and 1980s. Often towed from farm to farm by sugar beet haulage contractors they included machines made by Armer Salmon, CTM and Hestair Farm Equipment. The hopper on the 60 tons an hour rod-link Armer Salmon Hi-Vol model could be loaded from all three sides. Hydraulic rams, supplied with oil from the loader's own hydraulic unit, were used to set the elevator to a maximum 12 ft delivery height. The CTM Sample Master with a picking-off table made by Harpley Engineering was recommended for use with multi-row harvesters. The speed of a picking-off table with room for two people could be varied to suit conditions and in good going the Sample Master could load up to 40 tons an hour. Hestair Farm Equipment, which made the Todd Contractor loader, suggested that when using a forklift it was possible to load lorries at a rate of 1½ tons a minute.

The change from single-row trailed harvesters in the 1950s to high output six-row self-propelled machines in the 1990s was reflected in the development of cleaner loaders. Typical mid-1990s cleaner loaders, including those made by Armer Salmon, Harpley Engineering and Terry Johnson with an 8-20 hp

4.40 Sales literature explained that an output of 49 tons an hour was possible when filling the hopper on the Lister cleaner-loader with a root fork on a tractor front-end loader.

4.41 The three-point linkage-mounted John Salmon rod-link cleaner loader was driven from the tractor pto.

4.42 The CTM loader had a rotary cleaning cage.

diesel engine, had an output of 60-120 tons an hour. The 120 tons an hour Armer Salmon Hippo with a picking-off table driven by a hydraulic motor was big enough for two people. The CTM loader range, including the Brettenbridge 130B, with a 20 hp diesel engine and hydraulic motor drives to the barrel cleaning cage and delivery elevator, only took fifteen minutes to load a 30 ton lorry. Terry Johnson at Holbeach made the 120 tons an hour Todd cleaner loader with a Lister diesel engine in the early 1990s. The optional picking-off table was recommended when using a Todd loader on farms where the crop was grown on fields heavily infested with stones.

Chapter 5
Potato Harvesting Machinery

A flat-tined fork or a horse-drawn potato plough and occasionally a spinner or an elevator digger was used to lift the potato crop in the 1930s and 1940s. Grubbing around in the soil to find buried potatoes was all part of the day's work. Horse- and tractor-drawn potato ploughs, similar to a potato ridging plough with equal-sized wheels and prongs extending beyond the lifting body, were used to loosen the potato ridges. The prongs sieved away some of the soil but many potatoes were left buried in the ground. Slatted mouldboards left more of the crop on the surface but the hand-picking gang still needed to stir up the soil to find the buried potatoes. It was usual to cultivate the field after the crop was cleared to bring any buried potatoes to the surface for the pickers to collect.

Bamfords, Bamlett, Blackstone, Dening, Ransomes and others made horse- and tractor-drawn potato spinners in the 1930s. However, many farmers were either unable to afford the cost of a spinner or did not have the necessary horse or tractor power to pull one.

Ransomes made potato diggers at Ipswich for the best part of eighty years. The Ransomes No 12 spinner, described in their catalogue for 1910 as a potato digger, had a wide share that passed under the ridge. Tines on a land wheel-driven vertical rotor threw the contents of the ridge sideways on to cleared ground.

Spinners and Elevator Diggers

The spinner or elevator digger used to harvest potatoes was an improvement on the earlier potato plough. However, no matter how well the spinner or digger was adjusted some potatoes remained in the ground. A potato-digging machine made at Chelmsford in 1855 was probably the first recognisable potato spinner. The wooden-framed spinner had a share running under the ridge and a set of land wheel-driven wooden forks flicked the potatoes sideways on to the ground for the hand pickers.

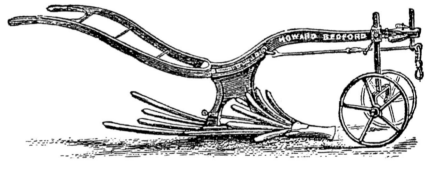

5.1 A pair of horses with a Howard potato plough lifted 3–4 acres of potatoes a day.

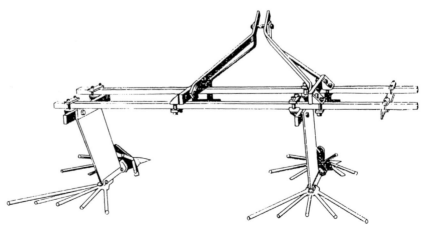

5.2 Alternate rows were lifted with the potato-raising bodies on a Ransomes mounted FR toolbar.

Design progressed from trailed land wheel-driven machines to pto-driven potato diggers mounted on the three-point linkage.

The No 12 digger was supplied either with fixed tines or cam-operated hanging tines arranged to remain in a vertical position on the tine wheel.

The No 21 and No 28 hanging tine diggers with the option of a front head wheel were current in the 1930s and 1940s. The trailed No 41 with a rope-operated self-lift clutch was the first

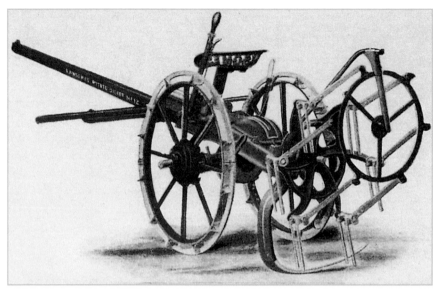

5.3 The Ransomes No 12 potato digger had cam-operated hanging tines.

5.4 Ransomes introduced the No. 41 pto-driven potato digger in 1946.

5.5 *The Ferguson potato spinner cost £84 in the mid-1950s.*

Ransomes pto-driven potato digger. The mounted FR digger, launched in 1947 and designed for the E27N Fordson Major, had a triple vee-belt drive from the power shaft to the digger gearbox. Later designated the FR PD4 it was superseded by the TPD (Tractor Potato Digger) 1008 in 1957. The TPD 1008 was modified several times during its twenty-year production run. Described in sales literature as simple to operate it had a safety slip clutch in the pto drive, twelve hanging tines and a side net to contain the potatoes within a narrow band.

There were two basic types of potato spinner in the early 1950s. Spinners made by Ransomes, Dening, Tamkin and others had a single vertical rotor rotating above a broad share running under the potato ridge while Bamford, David Brown, Ferguson and Lister Blackstone potato spinners had two semi-horizontal rotors. One rotor, which rotated in a clockwise direction above the share, moved the ridge sideways on to a second anti-clockwise rotor which removed much of the remaining soil to leave most of the potatoes on the surface for the hand pickers.

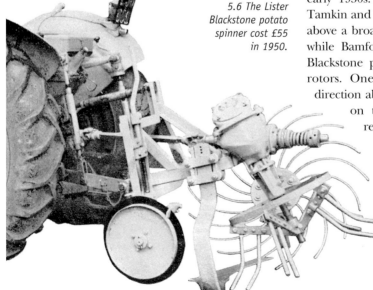

5.6 *The Lister Blackstone potato spinner cost £55 in 1950.*

Most spinners had a net or a canvas screen to deflect the potatoes in a narrow band. This helped to speed up hand picking but it was of little use if the share was set too deep. With large quantities of soil passing on to the second rotor many of the potatoes were covered and the pickers had to

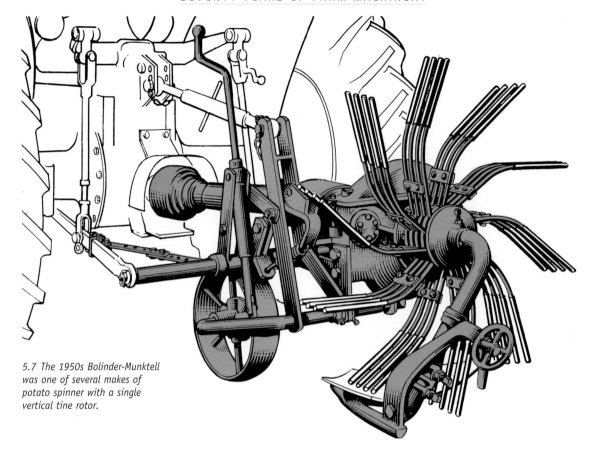

5.7 The 1950s Bolinder-Munktell was one of several makes of potato spinner with a single vertical tine rotor.

search for them in the soil. When the share was not deep enough some potatoes would be cut into two or more pieces.

Some farmers lifted potatoes with an elevator digger in the 1920s and 1930s. Known in some parts of the country as a hoover they were made by several companies including International Harvester, Massey-Harris and Oliver. Elevator diggers had a full-width share that lifted the complete ridge on to a rod-link conveyor running on cone-shaped rollers. Agitators shook much of the soil through the rod links back on to the ground and with most of the soil removed the potatoes fell on to a second rod-link conveyor for further cleaning. A deflector plate at the back dropped the potatoes in a narrow band, leaving enough space for the tractor wheels when lifting the next row to avoid having to wait while hand labour picked up the crop.

Early elevator diggers, pulled by either a tractor or a pair of horses, were chain driven from the large diameter land wheels. There were some exceptions, including the late 1930s International Harvester elevator digger with a petrol engine to drive the rod-link elevators. Some diggers had a gearbox to vary the speed of the rod-link conveyor in order to help shake away the soil when working on heavy land. A man riding on the digger lifted or lowered the digging share at the headland. Sets of different-sized agitators used to increase or decrease the shaking action of the rod-link conveyors were supplied with most elevator diggers.

Trailed and pto-driven Johnson and Peter Standen models and the Angus elevator digger supplied by Jack Olding were on the market in the late 1940s. By the early 1950s many farmers were using a semi-mounted or mounted one- or two-row elevator digger to lift their potatoes. The semi-mounted digger made by Johnson's Engineering at March was suitable for most popular makes of tractor. Carried on the lower hydraulic lift arms and pneumatic-tyred castor wheels at the back the Johnson single-row digger had work rates of 3-4 acres a day. The two-row model lifted 5-8 acres in a day. When lifting new potatoes farmers were advised to remove the rear

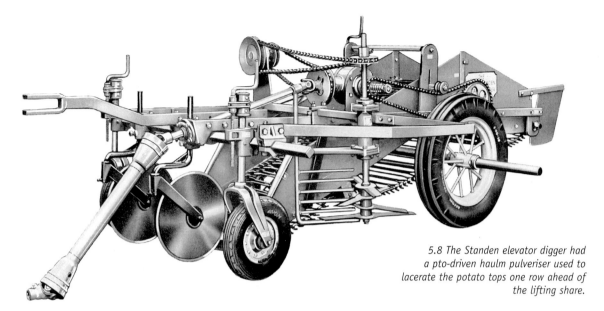

5.8 The Standen elevator digger had a pto-driven haulm pulveriser used to lacerate the potato tops one row ahead of the lifting share.

elevator to avoid undue damage to the crop. Pto-driven elevator diggers made by John Wallace & Sons in the late 1950s included single-row trailed machines with the choice of a rear- or side-delivery elevator, one- and two-row mounted models and a one-row semi-mounted digger.

Although a few complete harvesters had already appeared in the potato fields by the early 1950s most farmers were still using a spinner or an elevator digger. Some potato growers lifted the crop with a Wild-Bucher lifter or a Roulet Imp digger. MB Wild, which made the Wild-Thwaites field heap manure spreader, also imported the two-row Wild-Bucher swinging sieve potato lifter. It had two pto-driven swinging sieves directly behind a full-width share. A pair of wheels was used to set the working depth of the share. The

5.9 Lifting potatoes with a semi-mounted pto-driven Johnson elevator digger.

5.10 Picking potatoes behind an elevator digger.

5.11 Ransomes made one- and two-row elevator diggers.

contents of two potato ridges and, if necessary, the haulm passed on to the oscillating spring-tine sieves where most of the soil was removed and the potatoes were left on the surface for hand picking.

The semi-mounted Roulet Imp potato digger from Denmark had a scoop-shaped share which lifted the ridge on to a large diameter tined rotor. The soil fell between the rotor tines back to the ground and a fixed brush swept the potatoes off the rotor on to the ground.

The German Kluxmann potato lifter, introduced to UK farmers by J Gibbs Ltd at Bedfont in the early 1970s, could be used to harvest potatoes, vegetables and sugar beet. The Kluxmann lifted the entire ridge and passed the contents on to two oscillating grids running at right angles to the ridge. The vibrating steel tines on the grid removed the soil and left the potatoes in a narrow row for

5.12 Wild-Bucher swinging sieve potato lifters were made in Switzerland.

5.13 The Roulet Imp potato lifter was made in Denmark.

5.14 Sales literature explained that the Kluxmann root lifter, here seen harvesting leeks, held a seventy per cent share of the West German potato lifter market.

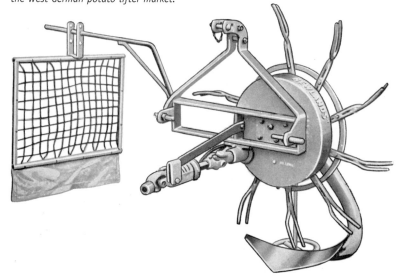

5.15 The early 1980s Newlands potato spinners were made at Linlithgow in Scotland.

hand picking. Although the choice was rather limited, small-scale potato growers were still able to buy a new spinner or elevator digger in the early 1980s. The Newlands pto-driven potato spinner, made at Linlithgow in Scotland, had a ten-fork reel driven through an oil bath reduction gearbox.

Complete Harvesters

The first potato harvesters were little more than an elevator digger with an extra elevator to lift the contents of the ridge on to a sorting table where hand labour picked off any stones, clods and weeds mixed up in the crop. Catchpole, John Salmon, Johnson's Engineering, Packman, Root Harvesters and Teagle were among the pioneers of mechanical potato harvesting. Most late 1940s and early 1950s pto-driven potato harvesters lifted the crop with a wide share which passed the contents of the ridge on to an open rod-link elevator. Most of the soil and stones fell through the links of the elevator web back to the ground while the potatoes were elevated to a hand-sorting platform. The potatoes were either bagged up or side elevated into a trailer running alongside the harvester.

Introduced as a prototype in 1945 the trailed Packman side-elevator harvester with an 8 hp air-cooled engine was an early example of a complete potato harvester. Awarded a silver medal at the 1948 Royal Show the Packman, with a haulm pulveriser in front of the lifting share, had an output of 2-4 acres a day.

Root Harvesters Ltd of Peterborough, formed in 1944 to manufacture potato harvesters, were making the Todd-Whitsed harvester in the early 1950s. Suitable for farmers with at least 30 acres of potatoes, the yellow side-elevator Todd-Whitsed harvester, named after Root Harvester directors Herbie Todd and Bill Whitsed, had an output of up to 4 acres a day. A 20 hp tractor was recommended for the mid-1950s semi-mounted bagger and side elevator Whitsed RB potato harvesters. The RB harvester was

Potato Harvesting Machinery

5.16 *The side elevator Whitsed RB potato harvester cost £598 in 1958.*

carried on the tractor's lower lift arms which were also used to put the share into and out of work at the headland. The standard RB harvester had steel rod-link elevator webs but for abrasive sandy soils some farmers bought the more expensive RB Continental model with the steel elevator links carried on endless rubber belts.

Johnson's Engineering made the Johnson Junior and the Johnson Major potato harvesters in the mid-1950s. The side elevator version of the Major cost £695 and the bagger model £710. Like most harvesters of the day the Johnson Major had a rubber-fingered 'hedgehog' conveyor to separate the stones and clods from the potatoes. A hydraulic ram to adjust the discharge height on the side-delivery elevator was an optional extra.

Catchpole Engineering, then the leading sugar beet harvester manufacturer, introduced a new type of potato harvester at the 1953 Royal Show. Instead of needing the usual team of people working on the sorting table the Catchpole Shotbolt had an 80 gallon water tank to separate the potatoes from the stones, clods and haulm. Claimed to lift, clean and wash 2-4 acres in a day, the Shotbolt used 200-500 gallons of water to harvest each acre of potatoes. Shares lifted the complete ridge which was conveyed through a powerful blast of air from a fan to remove the haulm and loose soil before the stones, clods and potatoes were dropped into the water tank. Sales literature explained that the potatoes would float on the surface while the stones and clods sank to the bottom. An endless chain-and-bucket system dredged up the stones and soil and tipped them over the side of the tank. Chain-operated rakes dragged the potatoes across the tank on to an open link conveyor where they were rinsed by a fine spray of water as they were elevated into a trailer. However, there was a flaw in the system because wet potatoes did not keep very well in the store!

Eleven harvesters, including machines made by John Salmon, Johnson, Teagle, Shanks and Whitsed, were demonstrated at an International Potato Harvester event held at Beverley in 1958. The Teagle Spudnick with a wide elevator digger share and rod-link elevator carried the crop up to the sorting platform with three endless web conveyors running side by side. Hand pickers transferred the potatoes from the centre conveyor on to either of the two side conveyors used to carry them to the bagging-off platform. Stones and clods left on the centre conveyor were returned to the ground. The John Salmon potato harvester lifted and elevated the crop on to a wide cross conveyor. Hand pickers took the potatoes off the cross conveyor and put them on to a second conveyor which carried them to the side-delivery elevator or bagging platform. The John Salmon bagger harvester cost £490, the side-elevator model £515.

The Shanks Featherbed potato picker was so called because in the late 1950s it was widely held by the general public that farmers were enjoying a featherbedded existence. The tractor-mounted Featherbed picker, which harvested up to 16 tons of potatoes in a day, was used after the crop had been lifted with an elevator digger. The Featherbed picker consisted of a pto-driven conveyor running at right angles to the row and an elevator to carry the potatoes up to the bagging-off platform. The two workers, lying face down on a Dunlopillo cushioned platform only a few inches above the ground, picked up the potatoes and put them on the cross conveyor. A third person working on the bagging-off platform filled and tied the sacks. The makers suggested that another worker should walk behind the machine to collect any potatoes left on the ground.

Robert H Crawford & Son at Frithville in Lincolnshire introduced a similar two-stage potato harvesting system with a work rate of half an acre an hour in 1962. The crop was lifted with a rotary screen-type digger with a cross conveyor that left six lifted rows of potatoes on the ground in a windrow. Six people kneeling on a tractor-drawn platform picked up the potatoes and put them on a cross conveyor which carried them to a side-delivery elevator. The crop was discharged into a trailer running alongside the machine.

Entries at the National Potato Harvester

5.17 An advertisement for the Shanks Featherbed Potato Picker explained that stones, clods and haulm separation were not a problem as the potatoes were hand picked.

demonstration held at Ormskirk in Lancashire in 1960 included machines made by Benedict, Catchpole, Gordon C Clarke, Massey Ferguson, Teagle, Johnson and John Salmon. The unmanned prototype Teagle Automatic harvester used a set of rubber suction cups to pick up the potatoes from stones and clods as they moved along the sorting conveyor. However, the makers suggested that one person needed to ride on the harvester to collect any potatoes missed by the suction cups. The GVC harvester designed for stony land was made by Gordon C Clarke at Malton in Yorkshire. An oscillating share lifted the crop on to three pairs of spider cleaning wheels which in turn carried the potatoes to the sorting table.

The Massey Ferguson MF711 based on a prototype harvester made by the National Institute of Agricultural Engineering was launched in time for the 1959 potato harvest. The MF711 with its enclosed gearbox drives and an absence of rod-link elevators was a relatively silent machine compared with other potato harvesters on the market. A concave disc lifted the complete ridge on to a tined cleaning rotor that returned most of the soil and stones to the ground. A vertical drum conveyor lifted the potatoes to a rotary sorting table from where they were either bagged up or

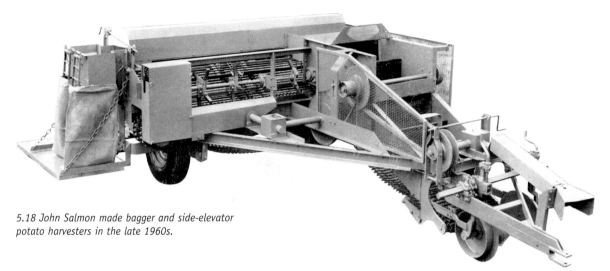

5.18 John Salmon made bagger and side-elevator potato harvesters in the late 1960s.

side elevated into an accompanying trailer. Daily maintenance was restricted to three grease nipples and only one spanner was needed to make the necessary field adjustments.

Undeterred by the lack of success with the Shotbolt water separation system Catchpole Engineering took the more conventional two-row self-propelled Shotbolt harvester to the 1960 Lancashire potato event. Built around a Nuffield Universal tractor without its front wheels the Shotbolt, with four hand pickers on board, recorded a work rate of about half an acre an hour. The Tubermatic, made by TSW Powell at Much Marcle in Herefordshire, used rubber-covered augers running above smooth rubber rollers to separate stones and clods from the potatoes. Another pair of rollers at the back of the machine got rid of the haulm.

5.19 Sales literature for the MF711 explained that potatoes had never before been handled with such kid glove treatment thanks to the harvester's rubber-covered components nursing them for most of their journey from soil to sack.

5.20 The two-row self-propelled Catchpole Shotbolt potato harvester.

Weimar Twin Master two-row harvesters with hydraulic depth control were imported by Leeford of London in the early 1960s. With a work rate of up to 7 acres a day the two-row Twin Master required a 30-40 hp tractor and up to five people on the sorting platform. The late 1960s semi-mounted Weimar harvester had eight cleaning stages with rubber and plastic cushioning which a sales leaflet explained was used lavishly throughout the machine to reduce tuber damage to an absolute minimum.

The German Grimme company, founded in 1861, made its first potato spinners in 1945 and a potato harvester production line was established in 1956 when Grimme built forty harvesters, thirty of which were sold

5.21 MB Wild made bagger and side-elevator potato versions of the Wild Firefly potato harvester in the mid-1960s.

5.22 Leefords of London imported the German Weimar Twin Master potato harvester in the early 1960s.

to Dutch farmers. The first single-row bagger and bunker Grimme VP harvesters, marketed in the UK as the Grimme Universal, were sold by Richard Pearson at Frieston near Boston in 1963. Pearsons made an alternative side elevator for the Universal harvester, which Grimme added to the list of optional equipment in 1964. The Universal had continental-type elevator webs and a rubber-fingered hedgehog clod separator. The single-row Gazelle bagger harvester, developed from the earlier Universal, was introduced in 1967 and the side-elevator and bunker models of the Grimme Gazelle remained in production until 1981. Other early Grimme harvesters included the Cadet for the smaller acreage grower, the two-row unmanned Continental and the Grimme GB for the UK market.

Catchpole Engineering introduced the two-row self-propelled side-elevator Fulcrop potato harvester in 1968. Unlike the earlier two-row Shotbolt built round a tractor, the prototype self-propelled Fulcrop, with a working speed of 1½ mph, had an 18 hp twin-cylinder Enfield diesel engine. A long elevator carried the contents of the ridge from the lifting shares to a pair of haulm stripping rollers. A mechanical soil-separating conveyor carried the crop on to a rotary sorting table. With a driver, six hand pickers on the sorting table and three tractor drivers with trailers the Fulcrop could, it was claimed, harvest up to 5 acres in a day.

Johnson's Engineering made the single-row Major, the J106 bagger, the J107 bunker and the J1067 combined bagger and bunker potato harvesters in the mid-1960s. The two-row Johnson Twin Major and the two-stage J202 six-row harvesting system with an output of 15-20 acres a day were being made at March when Ransomes bought Johnson's Engineering in 1968. The two-stage J202 six-row harvesting system consisted of three two-row semi-mounted elevator diggers and a pto-driven loader to collect the potatoes and elevate them into a trailer. The first J202 elevator digger lifted the two centre rows of potatoes and left them on the ground behind the machine. Two side-delivery diggers followed to lift the four outer rows; one delivered the crop to the right and the other to the left to form a windrow of six rows of potatoes ready for collection and loading into a trailer.

5.23 The Grimme VP harvester was marketed in the UK as the Grimme Universal.

Johnson's Engineering also imported Weimar and Faun potato harvesters in the mid-1960s. The German-built Weimar potato harvester range included one with a rotary sorting table and a side elevator. Another had two contra-rotating discs to lift the ridge and its contents on to rubber-covered conveyors and a side-delivery elevator loaded the potatoes into a trailer. The unsorted loads of potatoes, stones and clods were dealt with at the farm potato store. The share and rotary disc on the single-row rear-mounted Faun removed much of the soil before the crop was conveyed to the hand-sorting table. The potatoes were either side elevated into a trailer or conveyed to a bagging-off platform at the front of the tractor.

Ransomes had been making potato ploughs and diggers since the days of the horse when they bought Johnson's Engineering in 1968. The acquisition added Johnson, Weimar and Faun complete harvesters

5.24 The Johnson J202 harvesting system consisted of three semi-mounted elevator diggers which left six rows of potatoes in a single windrow.

5.25 The second stage of the Johnson J202 system used a pick-up loader to collect and load the windrowed potatoes.

together with Johnson elevator diggers, potato sorters and elevators to Ransomes' product range. When Bonhill Engineering became the UK distributor for Weimar harvesters in the mid-1970s Ransomes continued marketing Faun potato harvesters until Kverneland bought Underhaug Faun in 1986. Having lost the Faun franchise Ransomes introduced one- and two-row Hassia potato harvesters with hydraulic rear-wheel steering at the 1986 Royal Smithfield Show.

The Ransomes Johnson Monarch potato harvester for 50 hp-plus tractors appeared in 1971. The side delivery, bunker and bagger versions of the Monarch

5.26 The Faun potato harvester had an output of up to two acres a day.

5.27 The first Ransomes Monarch potato harvesters were made in 1970.

had a work rate of up to 4 acres a day. The Monarch bagging-off platform had a weigher and storage space for fifty sacks of potatoes. The Ransomes Sovereign replaced the Monarch in 1979. The digging share and elevator were offset from the tractor so that the tractor wheels ran on cleared land. There was sufficient room on the Sovereign's sorting table for eight pickers who worked under an all-weather canopy. Fifty Sovereign harvesters had been built at March when the line was discontinued in 1984.

5.28 The Whitsed Duplex harvester lifted up to 25 tons of potatoes in an eight-hour day.

Potato harvesters made by Root Harvesters at Peterborough in the late 1960s included the Whitsed Duplex and two-row Twin Duplex with large hand-sorting platforms and an optional kit for lifting carrots. Whitsed launched the unmanned Duplex Electronic harvester and the self-propelled Unitate harvesters in 1971. After five years of development work the makers claimed that the new X-ray sorting system on the Whitsed Duplex Electronic harvester had made hand sorters redundant. Its secret was the sixteen X-ray beams used to separate the stones and clods from the potatoes. This was done

5.29 The first self-propelled Grimme potato harvester was made in 1971.

by measuring the density of the clods, stones and potatoes with the X-ray beams as they dropped on to the conveyor. When the X-ray unit detected a denser stone or clod a row of seventeen rubber-sorting fingers on one side of the conveyor opened to return the unwanted stones or clods to the ground. The fingers, which remained closed when potatoes passed through the X-ray beam, were also used to deflect them on to a side-delivery elevator. Hand pickers were not completely redundant, however, as a seat was provided for a minder to pick off any potatoes missed by the X-ray beams.

The two-row self-propelled Whitsed Unitate with a Ford, David Brown, MF or International tractor skid unit, launched in 1971, harvested up to 5 acres a day. The Unitate had a built-in grading and crushing unit to smash up the clods and small potatoes before they were returned to the ground. The single-row Highlander and twin-row Victor and Vulcan harvesters with full hydraulic controls were current when Hestair acquired Root Harvesters in 1974. The X-ray sorting system was not used on the Unitate or the Hestair Root Harvester range but it was sold as a stationary sorting unit for use in farm potato stores.

A few self-propelled harvesters were at work in the UK in 1974 when the Barth, Grimme and the Krakei 422 imported by Hoekstra Trading at Boston in Lincolnshire were demonstrated at the Potato Marketing Board autumn demonstration.

The first Grimme self-propelled two-row potato harvester with a 100 hp Deutz diesel engine and hydrostatic transmission appeared in 1975. Advertised as a high-capacity harvester for difficult conditions the 30 cwt self-emptying bunker had a moving floor and an adjustable height unloading elevator. An earlier but less successful self-propelled Grimme harvester based on the trailed Commander and introduced in 1971 had a Massey Ferguson 165 tractor skid unit and its own hydraulic system.

Hoekstra Trading introduced the self-propelled two-row Krakei harvester with a Ford 4000 skid unit in 1972 and by the late 1970s Krakei harvesters had a Ford 5600 skid unit and a sixteen-speed gearbox. The driver, seated in a cab on top of the harvester, had a good view of the optional front haulm pulveriser. Hoekstra was still importing trailed Krakei Type 131 and self-propelled Type 641 potato harvesters in the early 1980s. The makers explained that although the trailed side-elevator Krakei was designed for unmanned operation it only took a few hours to install the optional sorting table.

5.30 Krakei self-propelled potato harvesters were made in Holland.

The two-row self-propelled Barth, made by Barth Drainage Machines at Mabelthorpe in Lincolnshire, was unmanned apart from the driver who sat alongside its low-slung engine. The crop was discharged into a trailer by a side-delivery elevator which was swung forward over the driving position for transport. Later harvesters marketed by Barth in the UK included the self-propelled four-row Barth-Holland Gigant with a 130 hp DAF diesel engine and ten forward gears and the two-row self-propelled Barth-Holland with the option of a Ford 5000 engine or a 115 hp Ford 2715 industrial diesel engine.

5.31 An optional automatic weighing attachment was available for the bagger model of the Standen Startrite potato harvester.

5.32 The Amac ZM-2 was also used to harvest other crops including onions, carrots, parsnips and flower bulbs.

Standen was one of the leading sugar beet harvester manufacturers when it introduced the Startrite and the Startrite Junior in 1972. Both were available with a bagging platform or side-delivery elevator and a rear elevator returned any haulm that found its way into the machine back to the ground. A 1976 advertisement for the Super Startrite and the Hereward side-elevator harvesters emphasised that an important design factor was their simplicity with no electronic or hydraulic hardware to 'act up' during the harvesting season. The bagger and bulk loading two-row Sterling, launched in 1982, was Standen's first self-propelled potato harvester. The Sterling with a 116 hp Perkins diesel and hydrostatic transmission could be set to work in 28-36 in row widths. A pre-topper was an optional extra.

Hestair Farm Equipment demonstrated the high-capacity unmanned Whitsed Phoenix automatic potato harvester with X-ray sorting and holding bunker at the 1982 potato harvester event. Hestair sold its agricultural machinery interests shortly after the demonstration and the Phoenix proved to be the last in the line of Whitsed potato harvesters. Other early 1980s machines included the Amac and Samro harvesters imported by Benedict Agricultural. The Dutch Amac trailed harvesters included the unmanned D1 single row, the D2 two row and the E2 two row with optional chat crushing equipment. The two-row self-propelled Amac ZM-2 and four-row ZM-4 had Same diesel engines, hydrostatic transmission and hydraulic variator speed pulleys for the elevator and side-delivery conveyors.

The Swiss-built Samro potato harvester range included the manned Major SC in bagger, bunker or side-elevator format, the Samro Spezial and the Samro Junior Spezial. The bagger and side-elevator Spezial, described as an unsophisticated machine for the smaller grower, harvested up to 3 acres a day with two or four people working on the sorting table. Except for a bigger bagging-off platform holding up to 15 cwt of potatoes the Junior Spezial was identical to the standard model.

5.33 A 70 hp tractor was needed for the unmanned Amac D2 potato harvester.

5.34 No more than four people were needed on the Samro Spezial sorting platform.

Grimme potato harvesters in the mid-1980s included bagger and bunker models of the two-row unmanned 'Q' Continental and GB and also the single-row Cavalier and Crusader with rubber 'hedgehog' clod and stone separator belts. Grimme publicity material explained that noise meant wear but the 'Q' Continental with rubber belt drives to most of the conveyor webs made it a remarkably quiet machine. A control panel in the tractor cab was used to operate the harvester's power steering and vary agitation on the main elevator web.

Standen returned to two-stage harvesting in the late 1980s. The first machine lifted two rows at a time and left the potatoes in a narrow band on cleared ground. After leaving the potatoes for a while to dry, a lifter-loader gathered the crop and elevated it into a trailer.

5.35 Grimme introduced the 'Q' Continental in 1983.

5.36 The Grimme RowOver launched in 1984, was a single-row offset harvester with the option of a bagger platform or a 1 ton self-emptying bunker hopper.

5.37 There was enough room for four people on the Kverneland Faun 2200 sorting table.

5.38 The mid-1990s Status 1750, made by Standen Engineering at Ely, had a full set of controls in the cab and another set on the rear platform. A TV monitoring system for the tractor cab was an optional extra.

5.39 The Grimme SF1700 had a Claas cab with fore and aft adjustment and five cameras to monitor the harvesting functions.

Following the acquisition of Underhaug Faun in 1986 Kverneland marketed the SuperFaun rear-mounted harvester together with side-elevator and bunker versions of the manned two-row Underhaug Faun 2200. Like most early 1990s potato harvesters they had mechanical stone and clod separation, hydraulic steering and all hydraulic functions were controlled from the tractor cab.

The single-row Grimme Governor launched in 1991 was typical of early 1990 potato harvesters. The Governor had an offset share that lifted the crop and passed it on to the first of three separator elevators. After final hand sorting the crop was held in a front-mounted bunker, weighed and packed in 25 kg bags on a bagging-off platform or side elevated into a trailer. Two-row Grimme harvesters in the mid-1990s included the self-propelled SF1700 with a six-cylinder 280 hp Mercedes diesel engine, an infinitely variable hydrostatic transmission, a rubber track on the rear left-hand side and a hydraulically driven front-mounted haulm pulveriser.

Haulm Pulverisers

Some potato growers cleared away the potato haulm by chemical or mechanical means before lifting the crop. Destroying the haulm also helped to control blight on fields where main crop potatoes were grown. One method, long since banned, was to spray the tops with sulphuric acid or other haulm-destroying chemicals about a week to ten days before lifting the crop. This treatment was also used to control the spread of potato blight.

Most mechanical haulm pulverisers consisted of pto-driven rotating knives or rubber flails mounted on a horizontal shaft. The blades were contoured to match the shape of the potato ridges and on some machines the blade formation could be varied to suit a range of row widths. The two-row mounted Howard haulm pulveriser made by Rotary Hoes had power-driven blades running at 600 rpm on flanges that could be positioned on the rotor shaft to suit a range of row widths. Some potato harvesters had a belt-driven haulm pulveriser unit attached to the machine to destroy the vegetation one row ahead of the lifting shares.

5.40 Sales literature for the early 1950s Pest Control Rotoflail explained that the high-speed rubber flails made a rapid job of disintegrating potato haulm.

5.41 Profiled knives on the two-row Johnson-Underhaug haulm pulveriser were shaped to follow the contours of the potato ridges.

5.42 The mid-1950s Howard potato haulm pulveriser advised bystanders to be aware of flying stones.

5.43 The flails could be set to match the ridge on the early 1980s Amac haulm pulveriser.

5.44 The late 1950s Maulden clamp coverer cut out the spadework when covering potato clamps. When used at 1½ miles an hour it was claimed to put three layers of soil on both sides of a clamp a quarter of a mile long in an hour.

Chapter 6
Estate Maintenance Machinery

Hedging and Ditching

Well into the twentieth century farm workers still spent many long hours during the winter months cutting hedges and digging out ditches with hooks and spades. Life on the farm changed in the late 1940s when instead of cutting hedges with a long-handled slasher some farmers used a tractor-mounted hedge cutter probably made by Blanch, Bomford or McConnel. However, farm workers were still able to keep warm at elevenses time by standing by a crackling bonfire of hedge trimmings.

Early hedge cutters with a heavy-duty cutter bar, driven by a stationary engine, pto shaft or belt pulley were attached to a frame above a tractor, mounted on the hydraulic linkage or attached to a front-end loader. By the early 1950s several companies,

beam. The tractor driver positioned the cutter bar to trim the sides and top of the hedge. The Bomford Hedgemaker, which appeared the following year, had a horizontal beam pivoted on a frame at the back of the tractor. Made by Bomford Brothers in Worcestershire the Hedgemaker, a 13 ft 6 in reach, had a petrol engine with a long flat belt to drive the 5 ft cutter bar.

The late 1950s Bomford Warwick Hedgemaker, with the cutter bar at one end of a floating beam, was pto-driven through a system of pulleys and belts. Hydraulic rams were used to set the cutting height and the angle of the cutter bar. The early 1960s Bomford Avon and Arrow cutter bar hedgers had a

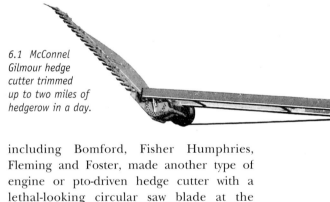

6.1 McConnel Gilmour hedge cutter trimmed up to two miles of hedgerow in a day.

including Bomford, Fisher Humphries, Fleming and Foster, made another type of engine or pto-driven hedge cutter with a lethal-looking circular saw blade at the business end.

The McConnel-Gilmour hedge cutter, introduced in 1948, was one of the first tractor-mounted mechanical hedge cutters. It had a centrally mounted horizontal beam carried on a frame above a Fordson Standard or similar tractor. A 4 ft 6 in long heavy-duty cutter bar at one end of the beam was driven by a long flat belt from a 2 hp air-cooled Petter engine on the opposite end of the

177

6.2 A man had to steer the Bomford Hedgemaker cutter bar along the hedge.

6.3 The horizontal boom and 5 ft cutter bar on the Blanch Model T2 hedge cutter gave it sufficient reach to trim most farm hedges.

Estate Maintenance Machinery

6.4 The Teagle Jetcut cost £40 in 1953.

new innovation in the shape of a 90 degree break-back mechanism. The Avon had hydraulic rams to set the cutting angle and height but spanners were needed for these adjustments on the cheaper Arrow hedge cutter.

A hedge-cutting attachment with a 5 ft cutter bar for front-end loaders, advertised by AB Blanch at Crudwell in the early 1950s, cost £101. The price included a Petter engine and two knives. Blanch also made two unconventional hedge cutters. The Blanch T2 cutter bar was attached to one end of a beam pivoted on a metal post bolted to the floor of a farm trailer. The Petter engine at the opposite end used to drive the cutter bar also helped to balance the beam. A man riding on the trailer had the job of manually controlling the position of the cutter bar as the trailer was pulled along the hedgerow. The cutter bar beam on the Blanch T4W hedger was attached to a post on a two-wheel trolley hitched to a tractor drawbar. A man riding on the hedger trolley steered with the foot controls and used a hand-operated winch to adjust the angle and height of the cutter bar.

A small hand-held cutter bar hedge trimmer was used on some farms in the late 1940s and early 1950s. Some were driven by a flexible cable drive from a garden tractor or a small generator-powered electric motor. Others had a hydraulic motor coupled to a tractor hydraulic system or a compressed air motor driven by a tractor-mounted compressor. The Teagle Jetcut hedge trimmer, with shoulder straps and a small two-stroke engine, cut growth up to ¾ in thick.

Sales literature described the self-propelled Baker & Hunt mechanical hedge trimmer with a small petrol engine as a machine that made easy work of a laborious task. Capable of cutting up to three years of growth, the flat belt-driven cutter bar was used to cut ditch banks, hedge sides to a height of 9 ft and the top of hedges up to 5 ft.

More efficient three-point linkage and power-driven cutter bar hedge cutters, including those made by Blanch, RJ Fleming, Allan Fuller, Parmiter and Twose, appeared in the late 1950s. The cutter bar on the rear-mounted Teagle Tracut, introduced in 1957, was belt driven from a pulley on the tractor

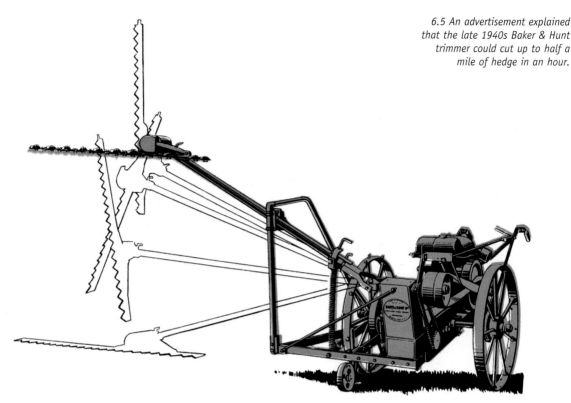

6.5 An advertisement explained that the late 1940s Baker & Hunt trimmer could cut up to half a mile of hedge in an hour.

pto shaft. Cutting height was set with a hydraulic ram but spanners were needed to alter the angle of the cutter bar. Hydraulic rams were used to adjust cutter bar height and angle on the Silver Bullet, added in 1963 to the Teagle range of mounted cutter bar hedgers.

Sales literature for the Parmiter mounted hedge trimmer with a 5 ft cutter bar, also launched in 1963, described it as a truly universal machine for any tractor with three-point linkage. Check chains or stabiliser bars were used to hold the machine rigid on the tractor. The Parmiter hedge trimmer was supplied with a plain knife and a serrated knife for cutting up to three years of growth.

Mounted hedge cutters made by Allan Fuller at Chepstow in the early 1960s included models with pto, engine or hydraulic motor drive. The mid-mounted Fuller Model 64 Tractordraulic hedge trimmer for many popular tractors, including the Fordson Major and Dexta, Nuffield 3 and 4 and the David Brown 950 and 990, was advertised as being able to cut up to five years of growth. The Fuller Tractordraulic had a hydraulic motor and roller chain drive for the knife crank and pitman, a relief valve protected the hydraulic motor from overload and a ram was used to set the angle of the cutter bar.

The McConnel Power Arm with a cutter bar hedge trimmer and other attachments, including a saw bench and concrete mixer, was one of the most popular estate maintenance machines in the early 1960s. The later rear-mounted and pto-driven McConnel Swingover hedger could, as the name suggests, be used on either side of the tractor. One of its advertised advantages stemmed from the fact that by swinging the cutter bar from one side of the tractor to the other the driver could cut a hedge while driving in either direction without the need to make any idle return runs. The hydraulic ram for the swingover mechanism was also used to set the cutting height while the angle of the cutter bar, adjustable through 360 degrees, was controlled with a hand wheel.

In the early 1950s a few hedge cutters with a circular saw blade came into use. Intended for cutting growth too thick for cutter bar trimmers, manufacturers included Fisher Humphries, Foster

Estate Maintenance Machinery

6.6 The cutter bar on the Fuller Tractordraulic hedge trimmer could be used to cut 10 ft high hedges.

6.7 A cutter bar hedger was one of the attachments for the McConnel Power Arm.

6.8 The man riding on the Fisher Humphries hedge cutter controlled the angle of the saw blade.

6.9 The Foster Rapidcut hedge cutter beam was turned through 90 degrees for transport.

6.10 A hydraulic ram was used to adjust the saw blade angle on the Taskers-Read hedge cutter.

and Stanhay. Growth up to 6 in thick could be cut with the trailed and pto-driven Fisher Humphries hedge cutter. The 36 in diameter saw blade on the Foster Rapidcut for Fordson, Ferguson and Nuffield tractors was awarded a silver medal at the 1949 Royal Show. The Rapidcut blade, driven from the pto through a system of shafts and belts, trimmed up to 800 yards of hedge in an hour. The saw blade on the Stanhay Hydrasaw hedge cutter was carried on a hinged frame between the front and rear wheels. A hydraulic motor was used to drive the saw blade and a hydraulic ram varied the cutting angle.

Sales literature for the early 1960s Taskers-Read hedge cutter explained that the 36 in saw blade, chain and vee-belt driven from the pto, could trim and fell up to fifty years of growth. The mid-mounted hedger was carried on a substantial frame and a heavy-duty mesh guard protected the driver from flying debris. The saw head was hinged forward and supported on a cradle for transport. A hydraulic ram was used to set the saw blade to its maximum cutting height of 12 ft and another ram adjusted the angle of the saw head.

Cutter bar and circular saw hedge cutters were included in the 1963 price list from RJ Fleming at Stony Stratford. Prices started at £144 7s for a hedge-cutting attachment for front-end loaders with a 5 ft cutter bar and 2 hp Petter engine. The most expensive Fleming hedge cutter with a 48 in diameter saw blade driven by a hydraulic motor was £499.

H Cameron Gardener, better known for the Rearloda, made mid-mounted circular saw hedge cutters in the early 1960s for Fordson Major and Power Major tractors. The 38 in diameter saw blade or alternative slasher blade was driven by a system of shafts and vee-belts from a bevel gearbox mounted on the tractor belt pulley shaft. A hydraulic ram adjusted the cutting height but spanners were needed to alter the angle of the saw blade.

Hydrocut at Sudbury made the world's first hydraulic motor-driven hedge cutter in 1953. The

6.11 A 42 in diameter saw blade or a 42 in wide flail head was used with the Fisher Humphries Lely Hedgeater.

new Super Hydrocut Mk III hedge cutter and scrub clearer, introduced in 1967, was a contractor's machine. The circular saw hedge cutter, mid-mounted on a heavy steel frame, was suitable for the Ford 5000 and similar tractors. A front-mounted 33 gallon/minute hydraulic pump supplied oil from a 40 gallon side-mounted tank to the 32 hp hydraulic saw motor. The Super Hydrocut cut hedges to a height of up to 13 ft 8 in and trimmed ditches up to 4 ft 6 in below ground level. Adjustment was provided to swivel the saw blade through 170 degrees. The MkIV Super Hydrocut, launched in 1970, was re-designed to meet new tractor safety cab regulations.

Flail hedge cutters made by Hydrocut, Fisher Humphries Lely, Twose and Teagle among others became popular in the mid-1960s. Public reaction to the appearance of hedgerows after they had been trimmed with a flail hedge cutter was initially one of horror and outrage. However, within a few years hedges cut regularly with a flail looked no different to those trimmed with a cutter bar machine.

According to sales literature the Fisher Humphries Lely Hedgeater flail rotor gobbled up growth of up to three years and there were no trimmings to clear up as the finely pulverised trimmings would soon rot away. The flail rotor was pto-driven through a gearbox and four vee-belts. As well as cutting the sides and top of a hedge the Hedgeater could also trim both sides of a ditch. The Fisher Humphries Lely Hedgehog with a saw blade, which was basically the same as the Hedgeater, was claimed to make light work of growth up to 12 in thick.

Three models of the DynaCut rear-mounted flail hedger were made by Teagle Machinery in the mid-1960s. The makers claimed that one man could attach or remove the DynaCut in a few minutes. The flail head was vee-belt driven from the pto and a hydraulic ram set the angle of the flail head. A cheaper version of the DynaCut had a manually adjusted flail head.

The McConnel Power Arm remained popular over the years and the PA44 Power Arm was current in the

Estate Maintenance Machinery

6.12 A hydraulic motor was used to drive the flail head on the Teagle DynaCut hedger.

6.13 The flail head on the late 1970s Bomford Bushwacker was driven by a hydraulic motor and rams were used to position the flail head.

6.14 Joystick controls and 50–75 hp flail motors were features of late 1990s Twose flail hedge trimmers.

6.15 The Foster Snitcher did most of the hard work when clearing up hedge trimmings.

mid-1970s. It was used for digging and trenching with buckets ranging from 5 to 39 in wide, ditch cleaning, handling manure, silage and grain with a variety of grabs and cutting hedges. Hedge cutting was done with a 39 or 48 in wide flail head while a selection of circular saw and slasher blades were used to deal with overgrown hedges.

Clearing and burning trimmings left by cutter bar and circular saw hedge cutters was still a job for hand labour on most farms. The Foster Snitcher, introduced in the late 1940s, provided a speedier alternative to clearing up trimmings by hand. Designed by a farm contractor both left- and right-handed Snitchers were made for the Ferguson TE20 tractor. The Snitcher, which looked like a side-mounted buck rake, gathered up the trimmings and left them in small heaps along the headland. The makers suggested that with a buck rake on the rear linkage the driver could gather up the small heaps of trimmings and carry them to a bonfire.

6.16 Four sizes of auger were available for the Massey Ferguson 723 post-hole digger.

Post-Hole Diggers

Digging holes for gate posts and for post and rail fences in heavy clay soil, even in mid-winter, was not a popular farm job. According to a 1950 sales leaflet the Ferguson post-hole digger for the TE20 tractor could, in most soils, dig a 12 in diameter post hole 3 ft deep in 30 seconds. It was explained that doing the job with a Ferguson post-hole digger was fourteen times faster than digging a hole by hand. The pto-driven free-swinging auger, which remained vertical even on unlevel ground, had a replaceable screw point and cutting blades.

Robot and RA Lister at Dursley both made post-hole diggers for most popular tractors. Three sizes of auger for boring 6, 9 or 12 in diameter holes were made for the early 1950s Robot post-hole digger which had a slip clutch and a safety shear bolt in the pto drive.

Holes up to 24 in diameter and 4 ft deep were dug with the mid-1960s Lister post-hole digger. The auger with straight-edged cutting plates for clay and gravel soils or serrated blades when working in hard ground was driven at 77 rpm through a double reduction gearbox. Clearing arms at the top of the auger shaft swept away loose soil to prevent it falling back into the hole.

6.17 The Robot post-hole digger was approved for use with Fordson Major tractors.

Ditching Machinery

Some farmers employed a contractor with a dragline excavator to clean out neglected ditches or dig new ones. Typical contractors' machines included the W&G Landrainer advertised in 1957 as a machine to dig or clean ditches and then level or load up the spoil. When digging a ditch 3 ft 6 in wide and deep the Landrainer with its own hydraulic system supplying oil to seven hydraulic rams had a work rate of up to a chain (22 yards) an hour. The early 1950s Whitlock Dinkum digger was another contractor's machine. Approved for the Fordson Major tractor and claimed to be the first tractor-mounted trencher and excavator, the Dinkum Digger was used by some farmers for drainage work, ditch cleaning and digging silage pits.

Several companies, including Barford, Bentall, Lawrence Edwards, McConnel and Stanhay, made various types of rear- and mid-mounted ditch cleaners and diggers in the 1950s and 1960s. The Cox ditch cleaner was one of several makes of rear-mounted ditcher. Supplied in kit form by Lawrence Edwards, the Cox ditcher consisted of a three-point linkage-mounted frame and a scraper blade to be used with a farm-owned front-end loader boom. A sales leaflet explained that it only took ten to fifteen minutes to put the frame on the tractor linkage, attach the loader boom to the frame and fit the scraper blade at the business end of the loader. After dropping the 3 ft wide chain-operated scraper blade into the ditch the tractor was driven forwards to drag the spoil from

6.18 The late 1950s Cox Ditch Cleaner was suitable for use with most popular makes of tractor.

6.19 The Stanhay Super Ditchmaster was awarded a bronze medal at the 1961 Royal Show.

6.20 As well as ditching and trenching, the mid-1960s Twose Multi-Scoop was also used to load farm yard manure.

the ditch. The Cox ditch cleaner with a 10 ft reach and 5 ft working depth was said to require no skill or special knowledge to clean ditches at rates of up to 3 chains in an hour. The early 1960s Becfab Ditcher/Dyker bucket on an 11 ft 6 in long rear-mounted jib was also used with the tractor backed up to the ditch.

The rear-mounted Stanhay Super Ditchmaster, with a rear-mounted jib and a pair of hydraulic legs to give it stability, was suitable for most popular makes of tractor. The Ditchmaster's hydraulically operated 3 cubic feet capacity grab, which slewed through 275 degrees, could clean or deepen ditches up to 11 ft deep and 10 ft wide. Optional attachments for the Ditchmaster included various grabs for loading sacks, bales, manure and other materials.

In 1953 the introduction of the tractor-mounted JCB Mk1 excavator, mainly for the construction industry, prompted the appearance of several rear-mounted farm ditchers. The early 1960s McConnel Power Arm, Barford Hydra-Ditcher, Twose Multi-Scoop and Bentall Hydra-Arm were all used with the tractor standing alongside the ditch. The bucket arm slewed through 180 degrees or more and several yards of ditch could be cleaned without moving the tractor.

A full range of attachments for the multi-purpose McConnel Power Arm, introduced in 1962, included a range of ditch-cleaning buckets. The No 12 bucket was said to make ditch cleaning cheap and easy because the bucket, which held a hundredweight of spoil, was force filled, raised and shaken empty with three hydraulic rams.

Suitable for most tractor three-point linkages the Bentall Hydra-Arm was used for digging or cleaning ditches and for offset trenching to a depth of 6 ft. The slewing mechanism enabled the operator to drop ditching spoil on to the bank or into a trailer. The Barford Hydra-Ditcher for tractors in the 45-60 hp bracket had its own self-contained hydraulic system and a 108 degree slewing arm. Attachments for the Hydra-Arm included trenching and sludge buckets, a manure fork, a hedge-cutting saw and a rotary cutter driven by a hydraulic motor for trimming ditch banks.

The early 1970s McConnel Power Arm 5 and 6 digger loaders with a twin-lever four function control box and a high seat for the operator had a reach in excess of 16 ft and a 12 ft digging depth. Attachments for the Power Arm 5 and 6 included a range of digging and trenching buckets and grabs for loading silage, sugar beet, manure, slurry and other bulk materials.

Other mid-1970s and early 1980s tractor-mounted equipment for farm ditch maintenance included David Brown, Horndraulic, MIL and Twose digger loaders. The Twose 190 with a 190 degree jib slewing angle dumped the spoil up to 11

6.21 The McConnel Power Arm 6 ditcher digger had its own hydraulic system with a 9 gallon oil reservoir and a pto-driven pump.

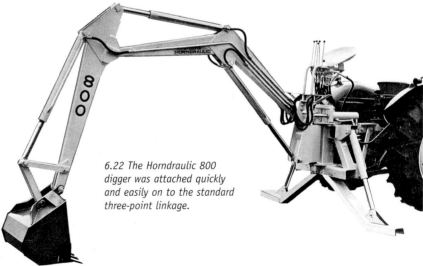

6.22 The Horndraulic 800 digger was attached quickly and easily on to the standard three-point linkage.

ft beyond the tractor wheels. The MIL Monarch and the Horndraulic 800 digger loaders with hydraulically operated stabiliser legs and multiple hydraulic control levers were used with various digging buckets and with grabs for loading manure, silage and sugar beet.

Some farmers cleaned and dug ditches in the 1980s and 1990s with a McConnel Power Arm or a similar rear-mounted digger ditcher. Others employed a contractor, who probably used a JCB backhoe digger and front loader.

Saws

Circular saw benches are almost as old as the steam engines first used to drive them. The Ransomes, Sims & Jefferies catalogue for 1886 included a range of circular saw benches with 36, 42 and 48 in diameter saw blades and an adjustable self-feed mechanism. Saw benches were driven from the belt pulley on a farm tractor in the early 1900s and several British companies made steel and cast iron-framed circular saw benches in the 1930s.

Bamfords at Uttoxeter made saw benches with a 24, 27 or 30 in diameter blade while drive from a tractor belt pulley was engaged with a fast and loose pulley arrangement. The late 1940s Bamford No 2 saw bench with a 24 in diameter blade cost £24 10s. Safety guards were not included in the price but sales literature pointed out that when the saw was used for trade purposes both top and bottom guards for the saw blade were required to comply with the Factory Acts. The saw blade shaft on the early 1950s Wrekin saw bench, made by James Clay at the Wrekin Foundry, ran on ball bearings and a removable section in the saw table facilitated blade replacement.

The introduction in the late 1940s of three-point

6.23 Bamford's No 2A saw bench was suitable for general farm use.

linkage-mounted and portable saw benches had the advantage of taking the saw to the wood instead of bringing the wood to the saw. Four sizes of the engine-driven Arun portable saw bench made by Walter A Wood at Horsham in Sussex in the early 1950s had 18-30 in diameter saw blades and 5-10 hp engines. McConnel mobile saws with 10-20 hp engines and a 30 or 36 in diameter blade were used for cross cutting and ripping timber. The saws were self-contained trailer units on large pneumatic tyres, making them easy to manhandle over rough ground and to tow at normal trailer speeds.

Two types of linkage-mounted saw bench were made in the late 1940s. One was really a stationary saw bench with a heavy cast-iron frame and three-point linkage hitch pins. The alternative three-point linkage-mounted Ferguson and similar cordwood saws with a swinging table were used mainly for sawing logs for firewood.

The pto-driven Dening of Chard Master mounted

saw bench for the E27N Major was sold with a 24 or 30 in diameter blade. A safety knockout clutch was built into the triple vee-belt drive. The Twose mounted saw bench with a 30 in diameter blade was suitable for most makes of tractor. Driven from the pto, its adjustable legs were used to level up the saw table when working on uneven ground.

The Ferguson Cordwood saw, made for Harry Ferguson by Robert Watson at Bolton, was belt driven from the tractor belt pulley. The belt was automatically tensioned when the saw was lowered on the three-point linkage and the spring-loaded swing sawing table returned to the start position at the end of each saw cut.

A similar rear-mounted belt-driven saw, made by McConnel for the Ferguson TE20, had a fixed table but the saw blade was offset from the tractor so that it could be used for cross cutting and for ripping long lengths of timber. The later three-point linkage-mounted McConnel All-Work saw bench also had an offset saw blade for ripping timber.

Chainsaws

The first chainsaw engines had a vertical float chamber carburettor, which only worked when the engine was upright. This limited chainsaw use to cross cutting with the guide bar held in vertical position. Early chainsaws had a long guide bar for the cutting chain which

6.24 The Arun portable saw bench was exhibited at the 1950 Smithfield Show.

6.25 Dening of Chard made the linkage-mounted Master transportable saw bench.

Estate Maintenance Machinery

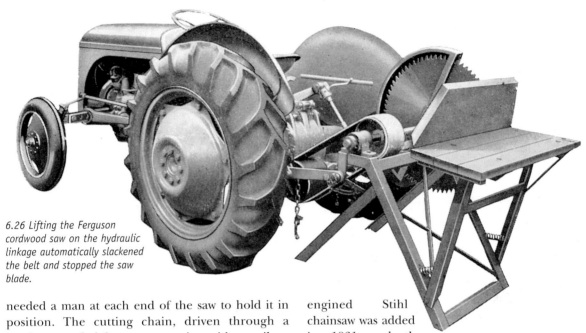

6.26 Lifting the Ferguson cordwood saw on the hydraulic linkage automatically slackened the belt and stopped the saw blade.

needed a man at each end of the saw to hold it in position. The cutting chain, driven through a gearbox, needed frequent attention with an oilcan until automatic chain oiling systems appeared in 1935. Lightweight one-man chainsaws were in common use by the mid-1950s, anti-vibration handles were introduced in 1965 and engines with electronic ignition appeared in 1968. The quick-stop chain brake was added in 1972 and the inertia-operated chain brake followed in 1982.

From the mid-1920s engineers in Germany, Sweden and Canada took a leading role in chainsaw development. Andreas Stihl, a German engineer, made a two-man cross-cut electric chainsaw in 1926. The first Sachs-Dolmar chainsaws were made in Germany in 1927 and a two-man Stihl chainsaw with a 6 hp petrol engine weighing a hefty 101 lb appeared the same year. An 8 hp two-man petrol-engined Stihl chainsaw was added in 1931 and the much lighter Stihl Lilliput saw weighing 55 lb appeared in 1933. The first Stihl one-man saw appeared in 1950 followed in 1954 by a really lightweight 4 hp Stihl chainsaw that weighed a mere 31 lb.

Introduced in 1941, the Danarm, the first British-made chainsaw, was designed by J Chubbley Armstrong and manufactured by TH&J Daniels at Stroud. It was a heavy two-man chainsaw with a 4ft long guide bar and a 250 cc Villiers two-stroke engine. Although the engine had a vertical float chamber carburettor, provision was made to rotate the gearbox and guide bar through 90 degrees to fell trees. The Danarm 28B chainsaw with a 7 ft long chain bar and a 350 cc Villiers engine appeared a

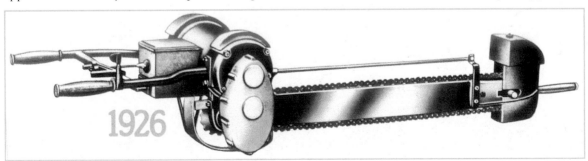

6.27 Stihl made its first two-man crosscut chain saws in 1926.

year or two later and the Danarm Junior chainsaw was added in 1945. The Junior's 1½ hp Villiers two-stroke engine, gearbox and chain guide bar were mounted in a two-wheeled tubular steel frame. The guide bar could also be rotated through 90 degrees to cross cut or fell timber. A detachable handle on the outer end of the guide bar enabled two people to use the saw when dealing with heavy timber.

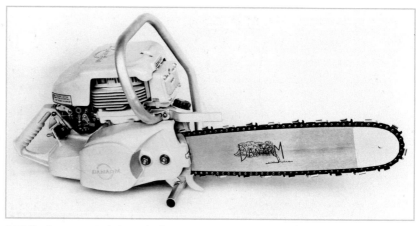

6.28 The Danarm 110 chainsaw had a 110 cc two-stroke Danarm engine.

Danarm, which made its first true one-man chainsaw in 1945, introduced the Danarm Tornado in 1946. The Tornado had a swivelling rear handle that allowed an operator to use the guide bar at different angles with the carburettor float chamber remaining upright. The Danarm Whipper, a new lightweight one-man saw, replaced the Tornado in the early 1950s. Advertised as a step forward in chainsaw design the Whipper had an 80 cc JAP engine with a recoil starter. The guide bar with an automatic chain lubrication system could be used for felling or cross cutting.

The new lightweight Danarm Fury chainsaw was advertised in 1956 as a more powerful successor to the famous Tornado. The Fury had a centrifugal clutch, a precision gearbox and a kick-proof rewind starter. The 23 or 30 in long guide bar could be rotated to cut at any angle and an optional handle was available to convert the Fury into a two-man saw. An extra long guide bar was available on special order. The Danarm DD8F chainsaw introduced in 1954 had a 98 cc Villiers 8F two-stroke engine with a Tillotson diaphragm carburettor that allowed the saw to be used at any angle. Weighing about 28 lb the DD8F, which was the first Danarm direct drive saw,

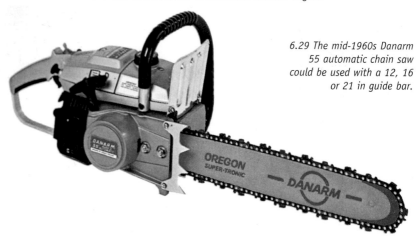

6.29 The mid-1960s Danarm 55 automatic chain saw could be used with a 12, 16 or 21 in guide bar.

remained in production until 1970. Other mid-1960s Danarm chainsaws included the Model 100 with a Danarm 100 cc engine and a range of guide bars up to 24 in long and the Model 110 with a 110 cc Danarm engine.

To meet a demand for smaller and lighter saws, three models of the Danarm 55 chainsaw with a 12, 16 or 21 in guide bar were current in the 1960s. The professional Danarm 55 had an anti-vibration handle and there were two versions of the basic 55 saw. One had the recoil starter on the left-hand side of the engine and the other on the right. An anti-vibration spring suspension system and a double baffle silencer for low noise levels were features of the Danarm 1-71-SS saw with a 71 cc Danarm engine introduced in 1973.

Danarm widened its chainsaw interests in 1979

6.30 A forty-year collection of chainsaws. Clockwise from bottom left: a 1950s vintage Teles Smith saw, a Sachs-Dolmar with a chain brake and anti-vibration handle made in 1991 and a 1970s Black & Decker electric hobbies saw. The 1970s Homelite XL has an automatic oiling system and the lightweight Sachs-Dolmar Oregon 91 saw made in 1992 with a 33 cc engine weighed a mere 8 lb.

when it became the UK distributor for Pioneer chainsaws made in Canada. This arrangement continued until Danarm saws were discontinued in 1984 when the company became the UK distributor for Zenoah chainsaws from Japan.

Buyers were spoilt for choice in the late 1950s when the British-made Danarm and Teles Smith chainsaws faced competition from several imported lightweight saws, including those made by Clinton, Sachs-Dolmar, Homelite, McCulloch, Pioneer, Remington and Stihl. Most of these saws had an automatic centrifugal clutch and either a direct or gear drive from the engine to the cutting chain.

Teles Smith made chainsaws from the 1940s to the early 1970s. Early models included a two-man saw with a 1·7 ft long guide bar and either a 3 hp petrol engine or a three-phase electric motor. Teles introduced the Little Tiger one-man saw, also with the option of a petrol engine or an electric motor, in 1956. The petrol-engined saw had an 80 cc JAP two-stroke power unit with a 'foolproof' float chamber carburettor and the 18 or 22 in guide bar head could be rotated for felling or logging. There were two models of the Little Tiger electric saw. A 15 amp, 3 pin socket was needed for the 240 volt saw and the 110 volt model was used with a mains transformer.

Early 1960s Teles Smith saws included electric models with 10-30 in long guide bars, the Little Tiger with a diaphragm carburettor and the Teles DD77 and the DD95 de luxe direct drive saws with Aspera engines. The later 5 hp D95 Super with 18, 22 or 28 in long guide bars had a Tillotson diaphragm carburettor and centrifugal clutch. A survey of chainsaws in 1968 found that the Teles Wizard with a 1 hp engine and 10 in long roller-nosed guide bar priced at £48 was the cheapest on the market.

EP Barrus imported Canadian Pioneer saws with 12-36 in long guide bars from the late 1950s and Trojan marketed American-built Clinton saws with 4 and 6 hp engines and 16-42 in long guide bars. The

6.31 The late 1990s Stihl 008 professional chainsaw with an 8.6 hp engine, a guide bar up to 48 in long and full state-of-the-art chainsaw technology dwarfs the 31 cc Stihl chainsaw for occasional use.

6 hp Stihl direct-drive Contra saw with a diaphragm carburettor and centrifugal chain clutch appeared in 1959. The mid-1960s Stihl range included saws with 13-33 in long guide bars and 55-140 cc engines. By the late 1960s some farmers were using a Stihl or other make of chainsaw with a special chain to cut blocks of silage from a pit or clamp.

Homelite saws, imported by Trojan from America in the late 1960s, had 60-100 cc engines and 12-42 in long guide bars. Homelite, which originally manufactured domestic electric generators, made its first chainsaws in 1949.

There were many makes and models of chainsaw available to the professional user by the late 1970s. Since then improved design has led to lighter and quieter saws with translucent fuel tanks and some even have heated handles. In the interests of safety the inertia quick-stop chain brake was added in the early 1980s. A typical 1990s farm chainsaw with electronic ignition and a 12 in long guide bar weighed about 12 lb.

Lifting and Carrying

Until the late 1940s most materials, including straw and animal feed, were moved around the farm by a horse and cart, on a wheelbarrow or carried on the stockman's shoulders. Farmyard manure was loaded with hand forks and carted to the field with a horse-drawn tumbrel or a tractor and trailer. Grain was bagged up on most combine harvesters and the coomb sacks were dropped on to the ground to be hand loaded on to a trailer. A coomb of barley weighed 2 cwt and a heavier coomb sack of wheat was 2¼ cwt.

Muscle power gradually gave way to mechanical lifting aids in the late 1940s. The three-point linkage-mounted Culverwell sack lifter, with a hydraulically lifted sack platform, did away with the heavy job of loading coomb sacks of corn by hand on to a trailer. Tractor-mounted hoists including the Whitlock Tracloder and Stanhay Ditchmaster, which doubled up as a sack hoist, provided an alternative way to load bales or sacks of grain. The late 1940s Whitlock Tracloder was bolted to the back of the tractor and a cable connected by a special linkage to the hydraulic lift arms raised the swivelling crane jib through 108 degrees. The Wild-Thwaites sack hoist on a trolley with a pair of pneumatic-tyred wheels was towed behind a farm trailer. A winch with 35 ft of wire rope and driven by a 2½ hp Villiers engine was used to haul sacks of grain up to the hoist and pull them up on to the trailer.

Harvest elevators for stacking hay, corn and straw date back to the early days of the threshing machine. Lister Blackstone and Salopian were still making

Estate Maintenance Machinery

6.32 The Whitlock Tracloder was used to load sacks, barrels and bales.

lightweight harvest elevators in the early 1950s when they were driven either by a flat belt from a threshing drum or a small stationary engine. These elevators were rather cumbersome for general farm use and were soon replaced by smaller petrol engine-driven sack and bale elevators including those made by Blanch, William Cook, Lister and Wolseley. The Wolseley Sheep Shearing Machine Co made two sizes of the Wolseley Universal elevator with a WDII petrol engine and delivery heights of 9 and 12 ft. The elevator was either hitched to a trailer for loading sacks and bales in the field or used to elevate hay into a stationary baler. The Cook and Blanch all-purpose elevators with 1½ hp Petter engines were used to load sacks and bales in the field or at the barn.

6.33 The Wild-Thwaites sack hoist could be used either behind or alongside a trailer.

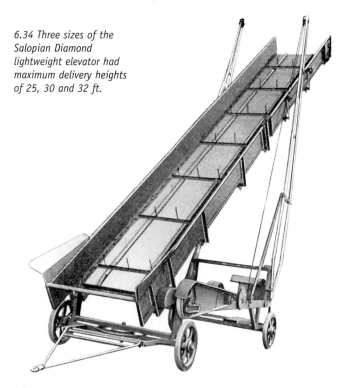

6.34 Three sizes of the Salopian Diamond lightweight elevator had maximum delivery heights of 25, 30 and 32 ft.

The growing popularity of tanker combines and specialist bale-handling equipment made the Wolseley and other sack elevators towed behind trailers obsolete. However, there was still a demand for multi-level engine-driven elevators with increased delivery heights for stacking bales in the barn. The 1960s and early 1970s British Lely Lelylevel elevator with a small petrol engine or an electric motor carried sacks, bales and boxes to a height of 20 ft. A pawl and ratchet jacking system on the Lister Mini and Major Multi-Level elevators made it possible to load sacks and bales either at ground level or from a trailer.

By the late 1950s the sack and bale elevators had been replaced on many farms by sack-, bale- and pallet-handling attachments for front- and rear-tractor loaders and tractor-mounted forklifts. Depending on the size of tractor the rear-mounted Catchpole tractor forklift,

6.35 The Wolseley Universal elevator was used for loading sacks, bales, sugar beet, farmyard manure and other materials.

Estate Maintenance Machinery

6.36 A sugar beet hopper was one of the attachments for the Blanch Universal elevator.

6.37 The late 1960s Lister Mini Multi-Level elevator was supplied with a 1¾ hp petrol engine or a ¾ hp electric motor.

6.38 Two or three 6 ft long sections of the mid-1960s Lister bale lift elevator with a 1¾ hp two-stroke engine could be joined together and used at an angle up to 45 degrees.

6.39 The hydraulic linkage provided the first lifting stage on the Catchpole forklift. An external ram on the mast frame completed the second stage of the lift.

6.40 A mast tilt ram was standard on the McConnel Tail Slave forklift but hydraulic side shift was an optional extra.

introduced in the early 1960s, had a maximum lift capacity up to 15 cwt. The Catchpole forklift had two pallet forks that could be spaced between 13 and 28 in apart. A tilt ram, which replaced the top link, was used to adjust the angle of the lift mast.

Other early 1960s rear-mounted forklifts included the McConnel Tail Slave originally made by Nu-Way Benson and those produced by Cameron Gardner and Curlight Industries. Most tractor-mounted forklifts had a side shift ram and lift capacities ranged from 1-1½ tons to a height of 10 ft 6 in or more.

Even more rear-mounted tractor forklifts with lift capacities of up to 30 cwt, including those made by Chieftain Forge, Cameron Gardner, Curlight, McConnel, Sanderson and Watveare, were on the market in the early 1970s. With more and more animal feed and fertilisers delivered on pallets or in big bags some farmers bought a self-propelled rough terrain fork truck. Bonser, Manitou, Sambron and Sanderson were among the first rough terrain forklift trucks to appear on UK farms. Most were two-wheel drive machines with a single mast and a lift capacity of 2-3 tons. Attachments included a pallet fork, a crane jib and buckets for corn, root crops, silage and manure.

Farmers had moved on to more powerful rough terrain forklift trucks with four-wheel drive, triplex (three-stage) masts and a cab by the late 1970s. Sambron two- and four-wheel drive rough terrain trucks with 45-72 hp diesel engines handled loads weighing up to 4 tons to a height of 12-16 ft.

Several companies, including Bonser, JCB, Manitou, Matbro, Sambron and Sanderson, were making rough terrain and telescopic fork trucks in the mid-1980s. Most could have two- or four-wheel drive and rough terrain trucks had duplex or triplex masts. Lifting capacities, depending on the size of truck, ranged from 1½-10 tons. JCB, which introduced its first telescopic handler in 1977, was

Estate Maintenance Machinery

making 2-5 ton capacity Loadall 3 ton rough terrain models with an optional four-stage mast. Pivot steering was a feature of the Matbro Teleram telescopic loader that lifted 2 tons to a maximum height of 18 ft. The Sanderson Teleporter range of telescopic handlers included a four-wheel drive model with four-wheel steering, an 86 hp diesel engine and power shuttle transmission.

Rough terrain trucks gradually lost out to the more versatile four-wheel drive telescopic handlers. By the early

6.41 Silage making with a late 1970s Manitou MB25P Farm Lift. It had a 65 hp Perkins engine, an eight-speed gearbox with hydraulic forward and reverse and a 3 ton maximum lift.

6.42 Telescopic telehandlers like this early 1990s 102 hp JCB Farm Special had hydrostatic transmission, front wheel, all-wheel and crab steering, joystick controls and a safe load monitor.

1990s many telehandlers had turbocharged engines in excess of 100 hp, hydrostatic transmissions, hydrostatic four-wheel and crab steering. Some had a pto shaft and trailer hitch. Attachments included pallet forks, grain and root buckets, manure forks, grabs and big bale spikes.

Farm Trailers

Horse-drawn tumbrels and four-wheel wagons with wooden wheels were the main form of farm transport in the 1930s and 1940s. Farm trials carried out in the late 1800s had shown that the draught required to pull a ton load on a four-wheel wagon was about 30 per cent less than the same load on a two-wheel cart. However, most farmers still used one-horse two-wheel carts for general haulage work and a two-horse farm wagon at haymaking and harvest time.

Robert Keen at Wallingford, Whitlock Brothers in Essex and many other local firms made horse-drawn carts and tractor trailers with pneumatic tyres in the 1940s. Local carpenters and blacksmiths were also called on to convert horse-drawn carts in order to hitch them to a tractor drawbar. Robert C Keen's price list for 1950 included a range of two- and four-wheel tractor trailers and the Keen commodious horse cart for medium to heavy horses. Suitable for heavy work on any type of land, the Commodious Horse Cart with an oak frame, a tongued and grooved elm floor, hay ladders and simple tipping device cost £63 10s. Second-hand axles and wheels with 7 in wide pneumatic tyres were used but for an extra £26 the cart would be supplied with a new axle, wheels and tyres.

There was an almost endless list of trailer makers in the early 1950s. The range of Keen North Stoke wagons at the 1950 Smithfield Show included 50 cwt tipping trailers and three-way hydraulic tippers. As most tractors of the day did not have a hydraulic system North Stoke tippers, finished with a coat of creosote, either had a screw-tipping mechanism or a hydraulic kit with a hand pump, oil tank and ram. Whitlock at Great Yeldham was making horse-drawn carts in the early 1900s. The 1948 Whitlock price list included fifty different carts and trailers, a few with steel sides but most made with timber. A catalogue informed potential buyers that the timber was soaked in a 1,000 gallon tank of red paint before assembly. Two- and four-wheel three-way Whitlock tippers and various horse-drawn wagons, tumbrels and water carts were made at Great Yeldham in the early 1950s.

6.43 The shafts on the Keen Commodious Horse cart could be substituted with a tractor drawbar.

6.44 The screw-operated tipping mechanism on this Whitlock steel-sided trailer had the capacity to tip a 3 ton load.

6.45 The Whitlock combined water cart and drinking trough with hinged top doors was made with steel or pneumatic-tyred wheels.

6.46 Whitlock farm trailers were finished with a coat of high gloss red paint; the floors were treated with creosote.

Other early farm trailer makers included JC Bamford, Salopian, Pettit, Taskers and Wheatley. The first Bamford two-wheel screw tipper was made in 1945, a four-wheeler was added in 1946 and within a couple of years hydraulic tipping trailers had been added to the Bamford range. In common with most early 1950s trailer manufacturers, the Wheatley range of trailers made at Peterborough included two-wheel fixed and tipper models, three-way tippers and four-wheel harvest wagons with harvest ladders and removable sides.

Harry Ferguson's two-wheel tipper and fixed trailers introduced in 1946 started a revolution in trailer design. With the wheels at the rear and the drawbar carried on the pick-up hook one third of the weight of the trailer was transferred on to the tractor. The 3 ton Ferguson trailer for the TE20 tractor made by James Sankey & Son at Wellington had wooden sides and floor. Optional equipment included over-run brakes, road springs, hay ladders and high sides used when silage making with a forage harvester. It was not long before other trailer manufacturers, including Pettit, Ransomes, Taskers, Weeks and Whitlock, were making 3 ton tippers with rear wheels and a combined ring hitch and drawbar clevis.

6.47. The harvest ladders or raves were an optional extra for the 6 ton Wheatley four-wheel trailer.

6.48 The 1950s Ferguson transport box was an easy way of carrying milk churns to the farm gate.

6.49 The mid-1960s Taskers front-mounted 4 cwt carrier box left the three-point hydraulic linkage free for other work. The box could also be used for front ballasting heavy rear-mounted implements.

A trailer was not always the most convenient means of carrying small loads around the farm. The alternative tractor-mounted transport box, popular in the 1950s, was used to carry small quantities of straw or animal feed and take milk churns to the farm gate. David Brown, Ferguson, Ransomes, Salopian and Taskers transport boxes carried loads up to 5 cwt, Whitlock made a much larger 15 cwt capacity box and Taskers made front-mounted 4 cwt carrier boxes. A wheelbarrow conversion kit with handlebars and a pair of pneumatic-tyred wheels was an optional extra for the Ferguson transport box. Fixed and tipping Ferguson transporter boxes with high sides were used to carry up to one cu yd of sand and other materials. Owners were advised to use front wheel weights when a heavy load was carried in the transporter box.

6.50 The 30 cwt Opperman Motocart, with rigid or screw tipper wooden body, had a top speed of 11 mph.

Estate Maintenance Machinery

6.51 Ransomes 3½ ton FR tipping trailers were made by FW Pettit at Moulton in Lincolnshire in the mid-1960s.

The three-wheeled Opperman Motocart and smaller self-propelled trucks including those made by Bonser and Wrigley were also used to cart sacks and bales around the farm. Introduced in the mid-1940s the 8 hp Motocart had an air-cooled engine mounted alongside the front wheel. A roller chain transmitted power to a four forward and one reverse speed gearbox and final reduction gears. It had rear wheel expanding shoe brakes and the steering wheel could be positioned for steering the Motocart either from the seat or while walking alongside the machine.

Martin-Markham, Massey Ferguson, Weeks, Wheatley and Whitlock were among the companies making 5 and 6 ton tipping trailers in the mid-1960s. Some had either timber or steel sides and floors, others were steel sided with wooden floors. Ten years later there were at least 200 models of farm trailer on the market. Farmers were offered an almost bewildering choice with two- and four-wheel drive, rear, three-way and high-lift tipping trailers made by more than twenty manufacturers.

Some four-wheeled trailers had a wheel at each corner while others had a tandem axle at the rear. Martin-Markham trailers, for example, ranged from

6.52 High sides for front- or side-loading grass when silage making with a forage harvester were available for the McCormick International 3 ton tipping trailer.

a 3½ ton tipper with optional silage sides and high sides for carting grain to 7 and 8 ton tandem axle hydraulic tippers with timber floors and high sides for carting silage, roots and grain. Scissor or high-lift trailers were also in fashion in the mid-1970s. As well as having a high tipping height for carting and tipping root crops these trailers provided a raised loading or working platform. Chutes in the tailboard were used for bulk filling grain drill and fertiliser spreader hoppers.

Farm trailers got even bigger in the early 1990s in order to handle the high outputs from combine harvesters, multi-row sugar beet harvesters and self-propelled potato harvesters. Small 2-5 ton tipper trailers were made along with 18 and 20 ton steel trailers on tandem axles to match the high horse power tractors used on big arable farms. Most of these trailers had a steel monocoque body with an automatic up-and-over opening tailgate, a sprung drawbar, hydraulic brakes and road lights.

6.53 Weeks high-lift trailers had two hydraulic rams, one raised the trailer sub-frame and the second tipped the body.

6.54 This 13 ton four-wheel tandem axle tipper with fixed sides was the largest AS Marston farm trailer in the early 1980s.

6.55 The Massey Ferguson 700 series of steel farm trailers ranged from a 4 ton drop side model to an 18 ton trailer with twin lift rams and an automatic opening tailgate with built-in grain chutes.

Index

Aktiv combine harvester 24, 34
Albion binder 11, 13
   combine harvester 22, 24
   cutter bar mower 89
   windrower 17
Allis-Chalmers 31, 40, 60, 101
   combine harvesters 20, 33
   forage harvesters 112, 113
   pick-up balers 60
   Roto-Baler 59
   tedder 101
Amac haulm pulveriser 175
   potato harvester 169
Armer Salmon cleaner loader 148
   sugar beet harvester 144
Armer sugar beet harvester 135, 144
ARM forage harvester 113
Arun portable saw bench 192
Axial flow combine harvesters 44

Baker & Hunt hedge cutter 179
Bale elevators 198
   handling 78
   loaders 80
   sledges 78
   throwers 67, 83
Balers 54
Bamford BM Volvo combine harvester 40
Bamford Claeys combine harvester 39

Bamford circular saw bench 191
   cutter bar mowers 88, 91, 93
   disc mower 96
   finger wheel rake 98
   forage harvester 119
   green crop loader 108, 110
   hay loader 106
   mid-mounted mower 93
   pick-up baler 69
   swath turner 98
   tedder 101
   Wizzler drum mower 95
   Wuffler 101
Bamford International 40
   pick-up baler 74
Bamlett cutter bar mower 87, 91
Barford bale loader 80
   Hydra-Ditcher 190
Barth potato harvester 198
Bean harvester 52
Becfab ditcher 189
Benedict Agricultural 96, 109
Bentall Air-o-Tedder 101
   Culti-rake 107
Best, Daniel 18
Big bale accumulators 85
Big balers 74
Binders 9
Birtley-Silk sugar beet harvester 131

Bissett binder 13
Blackstone tedder 97
Blanch, A B 38, 99
Blanch hedge cutter 179
    pick-up baler 64
    Whirlwind forage harvester 116
Blanch Lely bale elevator 197
    cutter bar mower 91
    finger wheel rake 99
    forage wagon 110
Blanch Snook green crop loader 109
BM-Volvo combine harvester 35, 40
Bolinder-Munktell combine harvester 31, 35
Bomford hedge cutter 177, 185
Bonhill combine harvester 43
    potato harvester 165
British Lely 38
    Lelylevel elevator 198
Brown Juggler bale sledge 82
Buckrakes 107
Busatis cutter bar mower 92

Cameron Gardener fork lift 200
    hedge cutter 183
Case combine harvester 20
    pick-up baler 73
Catchpole bale elevator 80
    cleaner loader 148
    fork lift 200
    potato harvester 159, 161, 163
    Powerbeet 146
    sugar beet harvester 131, 134, 140

swath conditioner 103
Caterpillar 18
Chainsaws 192
Circular saw benches 191
Claas big round baler 69, 75
    big square baler 78
    combine harvester 19, 22, 27, 38, 42, 45
    drum mower 95
    forage harvester 119, 121
    Huckepack 29
    pick-up baler 67
    trusser 58, 67, 69
Claeys combine harvester 38, 44
Clayson combine harvester 44
Clayton & Shuttleworth combine harvester 19
    cutter bar mower 86
Cleaner loaders 147
Colman & Co 38, 69
Combine harvesters 18
Cook bale elevator 197
    bale sledge 82
Cox ditch cleaner 188
Cracknell selective hoe 125
Crawford potato harvester 160
CTM cleaner loader 149
Culverwell sack lifter 196
Cutter bar mowers 86

Danarm chain saws 193
Dania combine harvester 35
    forage harvester 119, 121

Index

    pick-up baler 74
Dameco sugar beet harvester 129
David Brown 13, 89
    Albion combine harvester 24
    Hurricane forage harvester 114
    pick-up baler 64
    potato spinner 153
Dechentrieter combine harvester 22
Deering binder 10
    combine harvester 18
Dening circular saw bench 192
    cutter bar mower 91
    green crop loader 109
    Litedraft hay loader 106
    stationary baler 56
    swath turner 97
    tedder 100
Deutz-Fahr combine harvesters 6, 46
Disc mowers 94, 96
Ditchers 188
Doe silage combine 113
Double-chop forage harvesters 116
Down the row thinners 123
Dronningborg 74
Drum mowers 94
Dyson sugar beet harvester 141, 145

Elevators 196
Estate maintenance machinery 177
Europa Gehl forage harvester 118

Fahr combine harvester 36
    forage wagon 110
Farmhand bale accumulator 82
    big baler 75
    sugar beet harvester 141
Faun potato harvester 164, 173
Featherstone cutter bar mower 93
Ferguson cutter bar mower 90
    post hole digger 187
    potato spinner 153
    saw bench 191
    trailer 204
    transport box 206
Finger wheel rakes 98
Fisher Humphries combine harvester 38
    hedge cutter 181, 183
    forage harvester 82
    stationary baler 56
    sugar beet harvester 136
    threshing machine 14
Flail forage harvesters 114
Flail mowers 94
Fleming hedge cutter 183
FMC bean harvester 52
    pea viner 52
Forage boxes 110
    harvesters 112
    wagons 110
Ford combine harvester 30
Ford New Holland big square baler 78
Fordson sugar beet lifter 129

Forklifts 200
Fortschritt combine harvester 43
Foster Clarke sugar beet harvester 136
Foster threshing machine 14
Foster Rapidcut hedge cutter 183
    Snitcher 186
Fuller hedge cutter 180

Gappers 123
Garford sugar beet harvester 137, 144
Garnier pick-up baler 74
Gascoigne bale loader 83
GBW sugar beet harvester 137
Gehl forage harvester 118
Gleaner Baldwin combine harvester 33
Grain Marshall combine 23
Green crop loaders 107
Grimme potato harvesters 162, 167, 170, 173
Guyco sugar beet lifter 128

Harrison, McGregor & Guest 89
Hassia potato harvester 165
Hay conditioners 101
    loaders 105
    rakes 99
    sweeps 99, 105
Haymaking machinery 97
Haulm pulverisers 173
Hayter mower 94
Hedge cutters 177
Herriau sugar beet harvester 143, 147
Hesston big round baler 75

big square baler 77
    forage harvester 121
Holt, Benjamin 18
Homelite chainsaw 196
Hornsby trusser 54
Horndraulic ditch cleaner 190
Hosier green crop loader 109
Howard Big Baler 74, 77
    flail mower 95
    haulm pulveriser 173
Howard of Bedford 54
    stationary baler 55
Huckepack 29
Hudson sugar beet thinner 124
Hume pea swather 50
Hydrocut hedge cutter 184

International Harvester big round baler 76
    combine harvester 19, 41, 44
    sugar beet harvester 133

J C Bamford trailer 204
JCB digger loader 189
    telescopic handler 201
JF combine harvester 38
    Lightbinder 14
    swather 50
J Mann & Son 28
Jones Balers Ltd 60
Jones combine harvester 31
    forage harvester 119, 121
    pick-up baler 58, 60

Index

stationary balers 56
swath conditioner 103
tedder 101
John Deere combine harvester 20, 31, 46
    forage harvester 119
    pick-up baler 73
    sugar beet harvester 133
John Salmon cleaner loader 148
    sugar beet harvester 132, 141, 146
Johnson elevator digger 154
    haulm pulveriser 174
J 202 two-stage potato harvester 163
    potato harvester 159, 163
    sugar beet harvester 130, 137

Keen trailers 202
Kidd forage box 111
    forage harvester 121
    stripper harvester 17
Kluxmann potato lifter 156
Koela combine harvester 38
Krakei potato harvester 167
Kverneland potato harvester 165, 173

Lanz binder 12
    combine harvester 30
Laverda combine harvester 40
Lely bale accumulator 83
    bale elevator 198
    Baleightsom accumulator 83
    Cock Pheasant tedder 105
    combine harvester 38
    Cube-Eight bale accumulator 83
    finger wheel rake 98
    hedge cutter 184
Leverton 30
Lister Blackstone elevator 196
    hay rake 100
    potato spinner 153
    sugar beet lifter 131
Lister cleaner loader 149
    elevator 198
    post hole digger 187
    Take-Put bale carrier 80
Lorant 28
    trusser 58
Lundell 73
    forage harvester 118

Manea forage harvester 113
Manitou rough-terrain fork truck 201
Manned bale sledges 78
Marbeet sugar harvester 133
Marsh reaper 10
Marshall combine harvester 19, 23
    thresher 14
Martin-Markham Express forage harvester 116
    sugar beet thinner 125
    trailer 207
Massey Ferguson buckrake 107
    combine harvester 27, 42, 46
    cutter bar mower 90, 93
    disc mower 96
    forage harvester 114, 116, 118

pick-up baler 62, 74
potato harvester 160
tedder 104
trailer 209
Massey-Harris binder 13
combine harvester 19, 25
cutter bar mower 86
Dickie swath turner 98
hay loader 106
pick-up baler 62
potato elevator digger 154
stripper harvester 16
tedder 101
Matbro fork truck 200
Mather & Platt bean harvester 52
pea viner 51
Maulden potato clamp coverer 176
Maynard lifting wheels 140
McBain pea swather 49
McConnel bale packer 84
circular saw bench 192
hedge cutter 177, 180
mobile saw bench 191
Power Arm 91, 180, 186, 189
Tail Slave forklift 200
McConnel-Bomford stripper harvester 16
McConnel-Silk sugar beet harvester 131
McCormick, Cyrus Hall 9, 86
McCormick-Deering binder 13
cutter bar mower 90
combine harvester 18, 20, 23

pick-up baler 70
McCormick International bale loader 81
combine harvester 18, 23
cutter bar mower 90
forage harvester 112, 117
green crop loader 108
hay conditioner 103
pick-up baler 71, 74
McKay stripper harvester 16
Meijer bale sledge 82
Mid-mounted cutter bar mowers 92
Minneapolis Harvester Co 10
Minneapolis Moline combine harvester 20, 22
Minns sugar beet harvester 133
Mitchell-Colman forage harvester 113
Moreau sugar beet harvester 133, 143, 145, 147
Mounted cutter bar mowers 90
Murray sugar beet harvester 132

New Holland big round baler 75
big square baler 78
combine harvester 44, 46, 49
disc mower 96
flail mower 94
forage box 110
forage harvester 118
pick-up balers 58, 60
Stackliner bale wagon 83
swath conditioner 103
Nicholson hay conditioner 101
stationary baler 57

Index

swath turner 97
tedder 103

Oliver combine harvester 20
Opico 75
Oppel lifting wheels 140
Opperman forage harvester 112
    Motocart 206
    stationary baler 57

Packman potato harvester 158
Parmiter hedge cutter 180
Paterson buckrake 107, 112
Pea harvesters 49
Peg drum 18, 20
Pick-up balers 58
Pioneer chain saw 195
Post-hole diggers 187
Potato clamp coverer 176
    elevator diggers 154
    harvesters 158
    ploughs 151
    spinners 151
Potts baler 56
Precision chop forage harvesters 118

R A Lister 97, 99
Random bale sledges 79
Ransomes circular saw bench 191
    combine harvester 25, 31
    cutter bar mower 94
    hay maker 97

Hunter sugar beet harvester 146
pick-up trusser 58
potato digger 151
potato harvester 165
Power Beet harvester 146
stationary baler 54
sugar beet lifter 128
sugar beet harvester 141, 146
threshing machines 14
Tim sugar beet harvester 147
trailer 207
trusser 58
Reapers 9
Reaper threshers 18
Rear-mounted mowers 90
Rival pick-up baler 64
Robot-Hilleshog sugar beet harvester 131
Robot post hole digger 187
    sugar beet harvester 131, 136
Roerslev sugar beet topper and lifter 129, 131
Root Harvesters Ltd 148, 158, 166
Ross stationary baler 57
Rotary combine harvesters 44
Rotary mowers 94
Rotary tedder 105
Rotoscythe forage harvester 112
Rough-terrain fork trucks 85, 200
Roulet potato lifter 155
Rowcrop thinners 123
Ruston & Hornsby trusser 54

Sachs-Dolmar chain saw 193

Sail reapers 9
Sale Tilney 22
Salopian elevator 196
    Clearall hay loader 106
    sugar beet harvester 136
Sambron fork truck 200
Samro potato harvester 169
Sanderson Teleporter fork truck 201
Scott-Urschell sugar beet harvester 134
Self-propelled balers 58
    forage harvesters 119
    potato harvesters 163, 167
    sugar beet harvesters 145
Semi-mounted cutter bar mowers 89
Shanks Featherbed potato harvester 160
Shelbourne Reynolds stripper header 16
Shotbolt potato harvester 159, 161
Silage machinery 106
Silorator forage harvester 115
Six row sugar beet harvesters 147
Someca-Fiat disc mower 98
Standen elevator digger 154
    potato harvester 169
    sugar beet harvester 132, 137, 139
    Solobeet 145
Stanhay Ditchmaster 189, 196
    hedge cutter 183
    Windrum hay conditioner 101
Stanhay-Silk sugar beet thinner 123
Stationary balers 54
Steelfab Superbale bale collector 84
Stihl chain saws 193, 196

Stoll sugar beet harvester 139, 141
Stripper harvesters 16
Sugar beet cleaner loaders 148
    harvesters 125
    lifters 128
    self-propelled harvesters 145
    toppers 129
    top savers 131
    thinners 123
    three-stage harvesters 141, 143
    two-stage harvesters 129, 141
Sunshine binder 14
    stripper harvester 16
Sweeplift green crop loader 108

Tamkin potato spinner 153
    sugar beet lifter 128
Tasker buckrake 107, 112
    hay rake 101
    hedge cutter 183
    transport box 206
Taylor-Doe silage combine 113
Teagle hedge cutter 179, 184
    Spudnik potato harvester 159, 160
Teles Smith chainsaw 195
Threshing machines 14
Todd cleaner loader 148, 150
Todd Whitsed potato harvester 158
Trailers 202
Transport boxes 206
Trojan bale carrier 80
Tubermatic potato harvester 161

Index

Tullos Goodall Grass Conserver 102
    threshing machine 14
Twine knotter 10
Twose ditch cleaner 189, 191
    mounted saw bench 192
    sugar beet gapper 123
Twin-Mow rotary mower 94

Ugerlose forage harvester 118
Ursus Bizon combine harvester 43

Vendeuvre pick-up baler 65
Vicon Acrobat 99
    big square baler 77
    finger wheel rake 99
    Hippo forage wagon 110
    pendulum sugar beet thinner 125
    Monomat sugar beet thinner 125
    pick-up baler 74
    Rotorake 105
    Steketee sugar beet harvester 138, 141
Viking combine harvester 35

W & G Landrainer ditcher 188
Wallace binder 14
    elevator digger 155
    forage harvester 116
    stationary baler 56
Walley Gang-Mo 116
Warburton-Todd forage harvester 112
Warwick bale wagon 83
Watveare Overseas 36

Webb tedder 101
    Roloflo hay conditioner 102
Weeks trailer 208
Weimar potato harvester 162, 164
Welger big round baler 76
    pick-up baler 65
Wheatley trailer 204
Whitlock Tracloda sack hoist 196
    transport box 206
    trailers 204
Whitsed potato harvester 158, 166, 169
Wild Bucher potato lifter 155
Wild Firefly potato harvester 162
    forage harvester 116
    harvest thresher 16
Wilder Cutlift 106, 112
    forage harvester 115
    Speedi-Wilt 117
Wilder Steed green crop loader 110, 112
Wild Thwaites forage harvester 114
    sack hoist 196
Windrowers 17
Wire tying balers 55
Wolseley universal elevator 197
Wrekin saw bench 191
Wuffler 101

Old Pond
PUBLISHING LTD

Below is just a small selection of the wide range of agricultural books and DVDs that we publish.

For more information or a free illustrated catalogue please contact:

Old Pond Publishing Ltd
Dencora Business Centre,
36 White House Road, Ipswich,
Suffolk IP1 5LT United Kingdom

Website: www.oldpond.com

Tel: 01473 238200

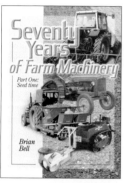

Seventy Years of Farm Machinery Part One: Seed Time
Brian Bell

Covering the period 1930-2000, Brian Bell deals with a wide range of machinery: tractors, ploughs, cultivators and harrows, grain and root drills, planters, manure spreaders, fertilizer distributors and crop sprayers. Hardback book

Seventy Years of Garden Machinery
Brian Bell

Covering the period 1920-1990, Brian Bell's book deals with the development of smaller machines: 2-wheeled garden tractors, rotary cultivators, 4-wheeled ride-on tractors, ploughs, drills, cultivators, sprayers, grass-cutting equipment, small trucks, miscellaneous estate items. Hardback book

Ransomes, Sims and Jefferies
Brian Bell

Ransomes invented the self-sharpening ploughshare and made, among other products, steam engines, lawn mowers, trolleybuses, threshers, reach trucks, tractors, subsoilers, disc harrows, sprayers, mowers, root-crop equipment and machinery for export. Brian Bell's comprehensive book emphasises 20th century farm machinery. Hardback book

Tractor Restoration: paintwork
Alan Davies

Professional restorer Alan Davies shows how he sets about preparing and spraying a tractor. He includes beating out damage, filling, stopping, spraying panels with primer and colour. He shows specific techniques for wheels and the chassis. The demonstrations are clear and detailed. DVD

Farm Machinery Film Records 1, 2 & 3
Brian Bell

This footage from the 1940s and '50s shows a wide range of farm machinery which was in development at the time. It comes from the archive of the National Institute of Agricultural Engineering and its Scottish counterpart. Altogether, over seventy machines are included in the programmes. Experimental prototypes from the NIAE itself are shown. DVDs

About the Author

Brian Bell MBE

A Norfolk farmer's son, Brian played a key role in developing agricultural education in Suffolk from the 1950s onwards. For many years he was vice-principal of the Otley Agricultural College having previously headed the agricultural engineering section. He established the annual 'Power in Action' demonstrations in which the latest farm machinery is put through its paces and he campaigned vigorously for improved farm safety, serving for many years on the Suffolk Farm Safety Committee. He is secretary of the Suffolk Farm Machinery Club. In 1993 he retired from Otley College and was created a Member of the Order of the British Empire for his services to agriculture. He is past secretary and chairman of the East Anglian branch of the Institution of Agricultural Engineers.

Brian's writing career began in 1963 with the publication of *Farm Machinery* in Cassell's 'Farm Books' series. In 1979 Farming Press published a new *Farm Machinery*, which is now in its fifth enlarged edition, with 35,000 copies sold. Brian's involvement with videos began in 1995 when he compiled and scripted *Classic Farm Machinery Vol 1*.

Brian Bell writes on machinery past and present for several specialist magazines. He lives in Suffolk with his wife Ivy. They have three sons.

Books and Videos by Brian Bell

Books in print
Farm Machinery 5th Edition
Fifty Years of Farm Tractors
Machinery for Horticulture (with Stewart Cousins)
Ransomes, Sims and Jefferies
Seventy Years of Farm Machinery: 1. Seedtime
Seventy Years of Farm Machinery: 2. Harvest
Seventy Years of Garden Machinery
The Tractor Ploughing Manual

DVDs
Acres of Change
Classic Combines
Classic Farm Machinery Vol. 1 1940-1970
Classic Farm Machinery Vol. 2 1970-1995
Classic Tractors
Farm Machinery Film Records Vol 1 Grain Grass and Silage
Farm Machinery Film Records Vol 2 Autumn Work and Rootcrops
Farm Machinery Film Records Vol 3 Testing and Prototypes
Harvest from Sickle to Satellite
Ploughs and Ploughing Techniques
Power of the Past
Reversible and Conventional Match Ploughing Skills
Steam at Strumpshaw
Thatcher's Harvest
Tracks Across the Field
Vintage Match Ploughing Skills
Vintage Garden Tractors